Understanding design and technology in primary schools

Cases from teachers' research

Edited by
Les Tickle

London and New York

First published 1996
by Routledge
11 New Fetter Lane, London EC4P 4EE

Simultaneously published in the USA and Canada
by Routledge
29 West 35th Street, New York, NY 10001

Routledge is an International Thomson Publishing Company I(T)P

© 1996 Selection and editorial matter Les Tickle;
individual chapters © their contributors

Typeset in Times by Florencetype Ltd,
Stoodleigh, Devon

Printed and bound in Great Britain by
Redwood Books, Trowbridge, Wiltshire

All rights reserved. No part of this book may be reprinted
or reproduced or utilized in any form or by any electronic,
mechanical, or other means, now known or hereafter
invented, including, photocopying and recording, or in any
information storage or retrieval system, without permission
in writing from the publishers.

British Library Cataloguing in Publication Data
A catalogue record for this book is available from the British Library

Library of Congress Cataloguing in Publication Data
Understanding design and technology in primary schools : cases from
   teachers' research/edited by Les Tickle.
      p.   cm.
   Includes bibliographical references and index.
   1. Engineering design–Study and teaching (Primary)–Case studies.
I. Tickle, Les.
TA174.U47   1996   95–45177
372.3′5–dc20   CIP

ISBN 0–415–13032–8

Understanding design and technology in primary schools

Teaching design and technology to young children has set new challenges for primary school managers, teachers, pupils and parents. Through the use of frank and detailed case studies, this book reveals the teaching aims and methods adopted by teachers, the issues they face in making their work effective, and the experiences of their pupils in learning design and technology. Extensive first-hand evidence of classroom experience is provided by the teachers.

The collection describes how action research can be done. It then provides practical examples of teachers introducing changes in the curriculum, in their teaching and in their use of evidence in monitoring teaching as a result of this kind of research. Student teachers, teachers, parents and curriculum managers will all benefit from the insights offered by this wealth of practical accounts.

Les Tickle is senior lecturer in education at the University of East Anglia. As a teacher, teacher educator and educational researcher, he has devoted over twenty years to improving the educational experiences of pupils and teachers. His combined interests in design and technology, the arts, the primary curriculum and teacher research are reflected in his previous books and are brought together in this one.

Contents

	List of figures	vii
	List of tables	viii
	Contributors	ix
	Acknowledgements	xi
1	**Design and technology in primary schools** *Les Tickle*	1
2	**Developing design and technology teaching through research** *Les Tickle*	10
3	**Combining design, technology and science?** *Martin Bayliss*	30
4	**Effective completion of technological tasks** *Sue Cooke*	38
5	**Mixed ability children and the single open-ended task** *Helen Deacon*	50
6	**Views and values** *Sarah Humphreys*	65
7	**Copying** *Rosemary Jackson*	77
8	**Change from rigid teaching** *Sue Lusted*	86
9	**Mental images and design drawing** *Andrew McCandlish*	95
10	**Working together** *Annette McMylor*	111
11	**Early years children, designers and partner choice** *Kevin O'Grady*	119

12	**Technology teaching at Dove First School** *Gillian Oliver*	131
13	**Children's choices** *Candy Rogers*	145
14	**Measuring success** *John Seaward*	156
15	**Food and design technology: where do we start?** *Val Simpson*	167
16	**'Seeing the light'** *Nancy Wright*	178
17	**Positive discrimination: is there a case?** *Dene Zarins*	201
	Index	213

Figures

4.1	The aspect of technology the children enjoyed most	45
4.2	How the children defined technology	45
4.3	The kinds of measures the children expected me to use in relation to their technology work	47
4.4	The evaluation criteria they thought I used	47
4.5	The children's evaluation of their own technology work	48
4.6	The evaluation they thought others would use	48
9.1	Examples of children's design drawings	97
13.1	Unexpected result – the theatre is made	152
15.1	Individual responses to open-ended tasks: the planning stage	172
17.1	Girls role playing using Playmobile, boys working with Lasy	205
17.2	George and Simon trying to get the motor going	206
17.3	Kyle is adjusting the drive machine, Sam is driving it	206
17.4	Sam and Simon driving their machines	207

Tables

5.1	Analysis of discussion, design stage	53
5.2	Analysis of discussion, prototype stage 1	54
5.3	Analysis of discussion, prototype stage 2	55
5.4	Analysis of discussion, making the wooden model	59
5.5	Analysis of discussion, adding mechanisms	61
5.6	Analysis of discussion, multi-storey model	62
5.7	A comparison of types of statements made by Hannah and Laura	62
7.1	Results of distance test	79
11.1	Infant children's perceptions of designers' attributes	126
12.1	Materials grid	133
12.2	Year R materials	135
12.3	Year 1 materials	135
12.4	Year 2/3 materials	136
12.5	New materials introduced at change of age range	136
12.6	Materials no longer available at change of age range	137
12.7	Year R skills	137
12.8	Year 1 skills	138
12.9	Year 2/3 skills	138
12.10	New skills at each age range	139
12.11	Skills teaching ceasing at each age range	139
12.12	Teacher confidence analysis scale	141
14.1	Feedback on self-evaluation	162
16.1	Learning methods and results of voting	186
16.2	Learning methods and results of voting	192
16.3	Summary of possible scores on worksheets	192
16.4	Recorded actual scores	193
16.5	Results of evaluation: learning methods (in order of points)	196
17.1	Number of boys and girls using the construction toys	203
17.2	The children's responses to the questions: Which of these toys do you have at home that you play with? Are these toys both girls' and boys' toys, boys' toys, or girls' toys?	209

Contributors

Martin Bayliss was coordinator for technology at Breydon (8–12) Middle School, and is now technology coordinator at Caister-on-Sea Middle School.

Sue Cooke is the headteacher (and class teacher) at Burston County Primary School.

Helen Deacon was the technology coordinator at Catton Grove Middle School when her research was completed.

Sarah Humphreys is the upper school unit leader at Burrowmoor Primary School, in March, Cambridgeshire. She was assessment coordinator at Dereham Church Middle School, East Dereham, when her research was completed.

Rosemary Jackson is a main scale teacher with responsibility for coordinating technology and French language teaching at Firside (8–12) Middle School.

Sue Lusted is the coordinator for children with special educational needs at North Denes (8–12) Middle School.

Andrew McCandlish is the deputy headteacher at Overstrand Voluntary Aided Primary School.

Annette McMylor is a Year 6 class teacher and technology coordinator at Wensum (8–12) Middle School. She was a Year 4 teacher when the research was done.

Kevin O'Grady is the headteacher at Roughton St Mary's (Endowed) Voluntary Aided Primary School. The chapter in this book was based on work done at Brooke County Primary School.

Gillian Oliver is a class teacher and technology coordinator at Dove First School, Harleston.

Candy Rogers is the deputy headteacher and teacher of reception class/ Year 1 at Parkers Voluntary Controlled Primary School, Saham Toney.

John Seaward is the headteacher at Lakenham Middle School, Norwich. He was the headteacher at Pulham Market Primary School when he did the research reported in this book.

Val Simpson is the deputy headteacher at Dereham Church Middle School, East Dereham. The chapter in this book is based on work done when she was a year group coordinator.

Les Tickle is a senior lecturer at the University of East Anglia.

Nancy Wright is a class teacher and technology coordinator at Fressingfield Primary School, Suffolk.

Dene Zarins is a teacher of Year 1 children at Avenue First School, Norwich.

Acknowledgements

I would like to thank colleagues in the School of Education and Professional Development at the University of East Anglia who have sustained the commitment to teacher- and school-development through the conduct of action research. It is a venture which is constantly reinvigorated by the willingness of very many teachers (and members of other professions) to engage in the challenges and risks of such research, and I want to record my gratitude to them all. This book is dedicated to all those teachers in primary schools who have fought against the odds to make the curriculum more practical and more intellectually challenging, through the introduction of design and technological activities. The odds, including lack of resources, inadequate training opportunities, and the imposition of particular curriculum content and assessment criteria, have been considerable. The dedication of those teachers to make the educational world of children an interesting place has remained commendable, and sustained their part in the struggle to improve the quality of schooling. I wish to acknowledge their work, and especially that of teachers who have made their work public for the benefit of others, including the contributors to this book.

Les Tickle

Chapter 1

Design and technology in primary schools

Les Tickle

Some of the first-hand experiences of teachers who have introduced design and technology into primary schools in recent years make up this book. The kinds of experiences they have encountered in their endeavours to effect design and technology teaching, and some of the practical issues which have arisen in doing so, are their central focus. The ways in which they have tackled problems through reflective and systematic thinking about curriculum implementation and development – their own research into their teaching and their coordinating responsibilities – provide the basis of the stories which are told. They convey the practical realities of teachers tackling the design and technology curriculum, the teaching strategies demanded by it, the collaboration with colleagues which it sometimes presumes, and the coordination of curriculum activities within a school. They also convey the kinds of problems and issues which have arisen for teachers, and the reflectiveness brought to bear upon those issues, in the search for understanding how best children might learn, and be taught.

Before introducing the individual stories, I want to set the scene for the teachers' chapters with a brief résumé of the development of design and technology as a primary school subject, and its changing nature during the recent period of innovation. This broader picture is important, because it displays something of the extent to which the challenges faced by teachers, in acquiring subject expertise and developing teaching strategies appropriate for young children, can be read as a collective professional problem. The later chapters, not surprisingly, do convey a sense of solitary activity and personal responsibility, and that is the way teaching often feels. But collectively, and set in the context of educational change, they become very evidently part of a shared response to the widespread experience of primary school teachers.

Design and technology for pupils in Key Stages One and Two in all primary schools in England and Wales is a legally enforced foundation subject of the national curriculum. In the early 1990s it was even described by some politicians as being the 'extended core' subject, putting its status almost on a par with maths, science and English. In one sense that is something of an amazing development, given that the subject came into the national policy framework for the primary curriculum only in the mid- to late-1980s. It was not, suddenly, a new subject which had been invented, but rather the elevation of a set of slow moving ideas to formal recognition. In another sense the amazement might be that, in a nation which led the way in industrialization and product development, design and technology took such a long, slow climb even to get in to primary schools. In that climb there were movements and shifts of direction within the subject itself, as it was formulated and remodelled.

The latest model is the result of the Dearing review (DfE 1995; SCAA 1995) which is described below, but it is worth putting that particular version into context. The value of doing so is, partly, to demonstrate that curriculum proposals, including those which resulted from the Dearing review, are the product of human interactions with all their strengths and frailties, ideals and compromises, powers and influences. Another purpose is to show that what we have in the current policy framework is a documentary model, a model whose relationship with the experiences of classroom life can be judged when reading the research reports which make up the following chapters of the book, or when thinking about one's own experiences in the classroom.

Both of these points are especially important in the case of design and technology. The detailed content and assessment criteria for each subject are established, within the framework of the Education Reform Act 1988 (amended by the Education Act 1993) as Statutory Orders. The power to amend these requirements lies with the Secretary of State for Education, and the nation was promised that the January 1995 Orders would not be changed for at least five years, in the interests of assuring some stability in the curriculum. In April 1995 however the Schools Curriculum and Assessment Authority published *Key Stages 1 and 2: Design and Technology the New Requirements* (SCAA 1995). Its opening statement indicates an acknowledgement of, and perhaps even infers a desire for, the internal dynamic of curriculum change in design and technology:

> In its present form, D&T is a relatively new and developing subject, changing rapidly both within and beyond education. The (statutory) requirements have been written in as non-prescriptive a way as possible, allowing teachers to decide how they will best be met in the light of their pupils' particular circumstances.
>
> (SCAA 1995: 1)

What we have is a documentary (legal) curriculum prescription with a (promised) five-year minimum life, imposed on teachers who are expected to be at least adaptive, and perhaps creative, in 'getting to grips with the new Order'. The attempt to reconcile legally enforceable prescription with the perceived dynamic of design and technological developments is no surprise, given the history of the subject and the school curriculum.

The background of learning, teaching and curriculum design which would fit the family of what we now call design and technological activities is summarized in Tickle (1987). It is largely one of the attempts to develop technical education in the secondary sector. There the tensions between academic study for entry into higher education, and practical studies for manufacturing occupations, were and still are acute (Goodson 1983). The universities and examination boards maintained the status of traditional academic subjects, to the detriment of more obviously 'vocational' ones. In primary schools the continued emphasis on 'the basics' of numeracy and literacy meant that even science teaching was in a parlous state as late as the 1980s. Design and technology had filtered down rather falteringly from the secondary sector (DES 1978; 1983; 1985a).

The final quarter of the twentieth century provides a picture of a dawning realization, at various levels in the education system, that the dearth of attention to technological education was a mistake. Curriculum development projects were sponsored to test out new kinds of activities or develop the more conventional craft syllabuses in schools. The schools inspectorate began to pay attention to the lack of practical and problem-solving dimensions of the curriculum. National monitoring of standards, from 1974 developed by the Assessment of Performance Unit of the Department for Education, included design and technology along with maths, science and English. This was indeed a new dawning, not only about national economic needs, but also about the supposed nature of so-called academic and practical curricula, and about the educational potentials of non-traditional subject content and activities for pupils.

I have described some details of the dawn chorus previously (in Tickle 1987: 6–28), in particular showing how craft, design and technology emerged through the mists of the 1980s. Its early versions as a national curriculum subject, in draft proposals and ministerial 'Orders', are rather complex, and need not be repeated here, except to indicate some of the consequent issues for teaching.

When it was eventually implemented within national curriculum legislation it had four attainment targets: Identifying Needs and Opportunities; Generating A Design Proposal; Planning and Making; and Appraising (i.e. evaluating). Tensions in defining specific content included the difficulties of teaching craft skills while allowing children to make decisions about the projects they would undertake. The place of home economics, business studies, and information technology remained problematic. Precisely which realms of scientific, technological, aesthetic, and technical skills content should be introduced at what stage was also hard to define. There was an intrinsic pedagogical problem of reconciling the acquisition of concepts with their application – how far might children learn concepts from making things, or make things without prior understanding in order to learn concepts from practical experience? Similar issues surrounded the question of how problem-solving processes could be pursued and developed, in association with the knowledge, skills and understanding which may be prerequisites to the conduct of problem solving.

Such questions were especially poignant for primary school practice. Attainment target one, which said that pupils should be capable of identifying and stating needs and opportunities for design and technological activities, through exploration of a range of design problems in the home, school and community, seemed so obviously 'out of order' in relation to young children. How could five and six year olds, for example, perceive such problems, appreciate possible solutions to them, and carry out the solutions? How could teachers of such young children encourage, respond to, accommodate and cater for a class full of individual design projects? Because of these and other difficulties arising from expectations placed on teachers and young children, curriculum 'Orders' were no sooner introduced than they were subject to revisions. The general turmoil, and sustained hope of a workable solution at the turn of the decade – into the 1990s – was summed up by John Eggleston:

> In 1989 Margaret Parkes's Technology Working Party report enjoyed more accolades than all the others put together. Visionary, revolutionary and inspiring were some of the adjectives used. Three years later technology education seems in crisis and the adjectives most often heard are demoralised, unworkable and even disastrous. . . . The pressure of teachers for a simpler, more practical and accessible subject has been shared by government, employers and many parents.
>
> (Eggleston 1992)

The simpler proposed version, published in December 1992, reduced the four attainment targets to two, Designing and Making, and formally introduced the mechanism of using designing and making tasks (DMTs, quickly to become known as dumpties), as what John Eggleston called 'the heart of Design and Technology'. Before these amendments were introduced as formal Orders to replace the first version, the whole of the national curriculum came into dispute, and the full review chaired by Sir Ron Dearing was established. A period of continuing dawn mist followed. The old Orders were still legally enforceable but overthrown by general acceptance of the new proposals. Yet the new proposals were subject to the same kind of review as the rest of the national curriculum subjects, in an intention to slim it all down and make it manageable. Some of the dawn mists evaporated in January 1995 with the publication of the details of the Dearing review, for implementation in September of that year, including the details for Design and Technology (DfE 1995).

It is in that context of innovation, rapid change, review, and modification that the research projects of these teachers were devised, and the chapters of this book written. Each results from a first and exploratory journey into the territory of teacher research. They were carried out in a range of different types of primary schools, with different age groups, dealing with different topics, and are written from the perspective of various kinds of teaching responsibilities. These chapters are organized alphabetically in the order of author surname. Teacher research as a means of developing teaching is a matter which I have also written about elsewhere (Tickle 1987; 1994). The ways in which the teachers did their research will be evident in each individual chapter, and will illustrate how the core ideas of teacher research were applied in each case. Some of its principles are set out in chapter two, in the form of a step-by-step practical guide, for readers who are interested in doing their own research.

Martin Bayliss provides an account of a technology coordinator's attempts to persuade colleagues to introduce design and technology teaching through the science curriculum, in order to solve the overcrowded timetable and gain a foothold in his school for cross-curricular teaching. His story is one of personal commitment to a particular set of ideas, which he was willing to introduce in his own classroom as a means of leading by example. His review of the innovation and of the premise on which it was based is a testament to the value of reflective thinking and willingness to take curriculum evaluation seriously, not only in relation to practice but to the aims, ideals and assumptions which underly the way the curriculum is organized.

Sue Cooke reports details of an investigation into the working of a group of children in a rural primary school, in which she sought to identify the reasons for their failure to complete their work, and/or the poor quality in the completion of it. It is also much more, for it records graphically the trials of a teaching head intent on carrying out research which will help her to understand the problem she had identified, and perhaps to solve it. The key elements of the research, though, relate to her Year 6 children's experiences of, and attitudes towards, the acquisition and use of 'making' skills, planning of projects, time allocation and management, and the use of criteria in judging what they make.

Helen Deacon examines how a single technology task set for the whole class, but with opportunities for different responses to suit individual needs built in to the nature of the task, is completed by a mixed ability group of ten year olds. The use of the single open-ended task, aimed to encourage children to work at their own levels of ability and allow them to develop and increase their personal skills, is evaluated through the question: why don't the products reflect the teachers' anticipated varying levels of design capabilities?

Sarah Humphreys records her exploration of the criteria that children use to evaluate commercially manufactured products, and products that they have designed and made themselves. This is a case study based on children's evaluations of electrically operated, working toys. It is also a record of the experiences of a novice teacher-investigator, struggling with the arguments, ideas and frustrations of teacher research, especially with the analysis of data which she collected from the activities and ideas of the children.

Rosemary Jackson summarizes an investigation of the phenomenon of pupils copying from each other in the classroom, and the reactions it provokes among children in Key Stage 2. The chapter defines different aspects of copying and their effects on, and perceptions by, children and teachers. An analysis of observation notes and interviews provides the basis for the development of policy and an action plan concerning copying in designing and making activities.

Sue Lusted describes in her own terms how she changed from a teaching style which required all children to produce identical objects, to an approach in which children were expected to use their own ideas, and in which they could achieve individuality in their products, within the guidance of a classroom project. It is an account of a self-reflective teacher concerned about the potential consequences of delegating decision making powers to pupils, about losing control of the quality of the work produced, and about adapting from previously 'proven' practices to new teaching strategies.

Andrew McCandlish studies children's technological drawings at the beginning of Key Stage 2. Classroom data is analysed to show how children regard their drawings, and the mental images used by them as starting points for recording their ideas. It provides a revealing study about the children's awareness of what they visualize, and what they draw, and the differences between these.

Annette McMylor describes her aim of encouraging cooperation among her pupils, and examines the apparent lack of a friendly, cooperative spirit in the classroom. The difference between her aspirations and what she perceived to be happening led to this study of how the children worked, who they worked with, and whether design and technology activities help towards fostering a more cooperative learning environment.

Kevin O'Grady provides another research reponse to the call for activities which involve children working in groups. He examined, through a study of Year 2 children, the criteria they use when selecting work partners for design and technology activities. His initial impressions that they did not choose work partners because of their perceived design and making capabilities were checked out against a range of evidence which had not previously been considered. Strategies of intervention which he sought to adopt for encouraging informed choices of partners are reported.

Gillian Oliver reports research which examined how she and colleagues in a first school began to teach design and technology, and how the author began to tackle her role as coordinator of the subject. That task was begun with an audit of the technology which was being taught, and in particular how resources, curriculum knowledge and teacher confidence affected what was being offered to the children. An agenda for change and development was created on the basis of the evaluation, and negotiated with the staff.

Candy Rogers shows how she thought she was providing opportunities for children to make choices in the reception class. She set out to observe and record the choices her children made when engaged in design and make tasks, and the reasons why they made them. Through observations and interviews the author was led to reflect on, and examine, the constraints she put on the children, and the reasons for them. A re-evaluation of teaching strategies led to changes in practice, and in the ways in which children's choices operated in the classroom.

John Seaward describes the ways in which a class of children nearing the end of Key Stage 2 use evaluation as a sub-process in their problem-solving activities. Classroom research was conducted to try to understand the capabilities of the children, and the chapter explores issues raised by the assessment of children against the 'appraising' criteria of national curriculum technology.

Val Simpson addresses the problem of the place of food in the technology curriculum. Strategies for teaching design and technology using food as a material are considered through the study of the gradual implementation of problem-solving activities, and pupils' responses to different modes of instruction and elements of teacher intervention. In her work she set out consciously to test three strategies, each with different measures of teacher control over decision making by pupils, and evaluated them alongside considerations of the dilemma of teaching knowledge and skills, whilst also encouraging individuality in the design of products.

Nancy Wright asked several questions about teaching and learning effectiveness, and used a class project on jointed movement to gather evidence related to the questions. The first was: how can the teacher ensure that effective learning is taking place? The second was: do the children have any idea what helps them to learn? The third was: is there any way that the teacher can find out what helps them learn best? This final question was the basis for a research plan, which in turn led to information and understanding on the others, by identifying when children gained new knowledge; what it was that helped them; and the effect of different teaching strategies in providing that help.

Dene Zarins considers whether there is a case for discrimination in favour of girls, in the use of construction toys and model making. Working with Year 1 children, her quest was to check whether forms of detrimental discrimination in learning opportunities occur within classroom activities, and whether deliberate intervention strategies to redress the balance can be justified.

Each of the chapters is based on the self-conscious selection of a particular research topic, sometimes based on a perceived problem, sometimes in curiosity about a phenomenon in the classroom, sometimes in a desire to monitor a new set of ideas and teaching strategies, or to evaluate the conduct of a new role. The features of learning and teaching which are represented are diverse, but still only form a small selection from the myriad of topics and research issues which have accompanied the introduction of design and technological activities in primary schools in recent years.

A notable feature of these events is that the organization of teaching and the choice and deployment of particular teaching methods is left to headteachers and teachers to determine. One of the characteristics of the research chapters is that it is these matters, and the nature of children's responses to learning activities, which have generated the teachers' research interests. This is a testament to their grappling with classroom issues which are much larger than the curriculum Orders themselves.

Those Orders, however, do provide another context for the teachers' work, and the activities which are presented to primary school pupils. Programmes of Study for Key Stages One and Two are pretty well consistent for children from age five through to eleven, with some indications of the kinds of advances in designing and making skills, and in knowledge and understanding, which are expected as they progress through school. According to the 1995 national curriculum documents (DfE 1995) designing and making skills are to be combined with knowledge and understanding, for purposes of formal instruction, and when pupils undertake tasks which involve designing and making products. This seems to me like a classic example of documentary curriculum statements which are presented to teachers to translate into pedagogic practices, with self-evident, inherent problems and dilemmas. It is the problems of translation to practice which the teachers expose and explore in their research reports.

FURTHER READING

Department of Education and Science (DES) (1978) *Primary Education in England*, London: HMSO.

Department of Education and Science (DES) (1982) *Education 5–9*, London: HMSO.

Department of Education and Science (DES) (1983) *9–13 Middle Schools*, London: HMSO.

Department of Education and Science (DES) (1985a) *Education 8–12 in Combined and Middle Schools*, London: HMSO.

Department of Education and Science (DES) (1985b) *Better Schools*, London: HMSO.

Department for Education (DfE) (1995) *Design and Technology in the National Curriculum*, London: HMSO.
Eggleston, J. (1992) 'Rebuilt better than new', *The Times Educational Supplement*, 18th December 1992.
Goodson, I. F. (1983) *School Subjects and Curriculum Change*, London: Croom Helm.
Jarvis, T. (1993) *Teaching Design and Technology in the Primary School*, London: Routledge.
Makiya, H. and Rogers, M. (1993) *Design and Technology in the Primary School*, London: Routledge.
Schools Curriculum and Assessment Authority (SCAA) (1995) *Key Stages 1 and 2: Design and Technology the New Requirements*, London: SCAA.
Tickle, L. (1987) *Learning Teaching, Teaching Teaching*, Lewes: Falmer Press.
Tickle, L. (1990) *Design and Technology in Primary School Classrooms: Developing Teachers' Perspectives and Practices*, Lewes: Falmer Press.
Tickle, L. (1994) *The Induction of New Teachers: Reflective Professional Practice*, London: Cassell.

Chapter 2

Developing design and technology teaching through research

Les Tickle

Some approaches to professional development actively encourage teachers to identify their own and their institutions' needs and to work out their own strategies towards meeting those needs, by way of research. The emphasis is on the identification and sharing of practical curricular concerns, gathering evidence to help to illuminate events within school practice, developing interpretations of the evidence in order to improve one's own and/or colleagues' understanding of what is happening in the classroom and school, and using expertise in the analysis of teaching to arrive at solutions which are based soundly on evidence. In this kind of school based professional development, higher education institutions are sometimes involved, providing support for teachers to research and to develop aspects of teaching and learning within their own schools. The precise issues to be investigated are determined by teachers themselves, according to their own most pressing professional concerns, but perhaps within the overall plans which their institution has articulated for its continued development. Tutorial support is intended to guide decisions about the focus of inquiry, its subsequent methods and conduct, and in the writing of reports which can be of value to others in the schools. The teachers' chapters in this book were based on this kind of approach.

In this particular case – the introduction of primary design and technology in the national curriculum – there are some general aims, issues, and content within which the focus of teacher research would obviously fit. But the professional development aims of such research are to develop confidence and capability in the knowledge, skills, understanding and practice of design and technology teaching. Where appropriate they may also be to develop curriculum coordination and leadership skills in order to provide for curriculum and staff development within a school. The broader context of innovation, and these professional development aims, open up a number of possible avenues for research and action which can contribute to one's own development, that of colleagues, and to our general understanding of what is happening 'on the ground'.

For example: for coordinators, research into the kind and extent of teachers' subject knowledge, the ways in which it can be developed, and how the knowledge held by teachers is transmitted to pupils, are all valuable topics for research which could lead to appropriate action in professional development provision in school. Issues of coordination, colleagiality, and leadership also arise when contemplating practical action for staff development. The development of strategies for effecting change depend upon the receptiveness of colleagues, and it is important to understand the conditions and processes which affect such receptiveness. Understanding the ways in which ideas are adapted and accommodated as relevant to particular children or working contexts is also important for effecting change. Such matters involve negotiation, and raise issues of autonomy, the delegation of responsibility, and matters of self-esteem, self-respect, and valuing of individuals within negotiations for staff development. All of these avenues of action might be effected more successfully on the basis of case study research.

These are matters which are likey to arise as practical action is contemplated and strategies are worked out for identifying staff development needs; for undertaking staff development planning; for negotiating provision to meet individual staff needs; for providing subject knowledge and teaching expertise in the transmission of that knowledge to pupils; and for engaging in curriculum planning for a subject within the context of school development planning.

It is possible, therefore, to think of a research project as one which begins by seeking data in order to understand a situation better, and thus to be in a position to make more informed judgements about how to proceed in staff development activities – an evaluation of an existing situation, leading to action.

It is also possible to embark on development activities, on the best available evidence, and to monitor one's own actions and the actions and responses of colleagues – an action research approach. Either way, a research project can be seen as one which is of value to one's own development of understanding, to the development of colleagues' subject knowledge and teaching skills, and thus as making a contribution to a school's development.

Similar options exist for the individual teacher within her or his own classroom. An action research plan might involve trialling new teaching aims and strategies, and gathering evidence about one's own actions and pupils' responses to them. The introduction of new learning materials or equipment, or the teaching of specific concepts or skills to particular groups of children, might be subjects which demand the gathering of extensive evidence on which to judge the quality of the learning which arises.

One's own teaching decisions and the values which underlie teaching aims, or the reasons why a particular innovation is being tried, may be important starting points for the design of the research. The search to discover how/why aims are being realized, or not, may be another.

Case study research might arise from a realization that all is not as one might hope in the pupils' learning. Several of the chapters in the book are of that type – pupils not completing work adequately, not working cooperatively, and so on. Research may be prompted by a simple puzzle – why do children produce similar work?, how do they feel about copying each others' work?, are the national curriculum targets appropriate for these particular children?

There are many possible starting points, so many that a great deal of what happens in schools is based on the evidence which is immediately (and impressionistically) available in the circumstances relating to each problem. Often the circumstances prevent the best use of evidence, or even cause the evidence to go unnoticed. A major objective of teacher research is to enhance the way in which evidence is handled, and hence how judgements are made in teaching. The following guidance has been written for teachers who have no previous experience of doing research. It has been based mainly on principles and methods of action research. It includes specific tasks which will help research projects to get under way, and references for further reading.

This does not preclude the possiblity of doing other kinds of research, which do not necessarily stem from or lead directly and immediately to action. For instance, case study of classroom and school processes, evaluation of the curriculum, historical research, or policy analysis provide valuable insights into educational activities which are the context for teaching in the classroom. Each of these other kinds of research may develop insight and understanding, through their particular modes of inquiry and scholarship, of factors which have a strong bearing on practical action. The research approach which is used may vary according to the nature of different issues or problems, and where 'in the system' they are thought to be located. Those differences can be considered more fully by further reading from the list provided later in this chapter, and will also be evident in the teacher research chapters which follow.

For the moment, it will be sufficient to note that one main difference between action research and other kinds of teacher research is that action research places the researcher at the centre of the research evidence: as one of the providers and as the gatherer/interpreter of it. In case study, or in the other kinds of research, the providers of evidence may be children, parents, other teachers, or sources such as documents. These may have a bearing on one's teaching role, but the practice itself is not necessarily the focus, and the rigour of self-reflection about one's own values, aims and actions may be less acute. This relationship will become evident as readers consider the following guidance, and especially if you begin your own research, because it will be necessary to decide where one stands in relation to the problem and the evidence.

In the examples of later chapters, those of Martin Bayliss, Sue Lusted, Candy Rogers, Val Simpson and Nancy Wright, for example, are in the action research mode. That of Andrew McCandlish is more clearly a case study of pupils' ideas, as is Rosemary Jackson's study of copying. Gillian Oliver's work is a form of whole-school evaluative case study. The differences are sometimes only subtle shades, but the importance of them in terms of the closeness to one's own ideas and actions, the sources of data, and the ways data are interpreted, are rather important.

BEGINNING A RESEARCH PROJECT

The purpose of any investigation is to provide a closer look at a chosen aspect of education. It may be a puzzle worth doing; a problem which needs solving; information worth gathering, or an issue worth addressing (by issue I mean a subject worthy of discussion or debate, one containing elements of dispute or alternative viewpoints and possibilities). Most important of all is to be able to focus on the nature and impact of events which occur within the workplace, in this case in design and technology teaching in primary schools.

To describe the situation in which one works, the aims one is pursuing, how one teaches, and why, describing for others what one already knows or what one would like to improve on in teaching may make one more conscious of some features of teaching which are otherwise taken for granted. Research will also make more sense to others when seen in a descriptive context. Writing in a descriptive way on these matters helps to reveal some important evidence, and also focuses the mind upon the task of selecting the most pertinent information from the morass of data which inhabit teaching situations.

Deciding the topic; judging the scope of the project; devising principles for the conduct of the research; deciding what data to select and collect, and from whom; and working out how to record it, will all exercise the mind from the start of a research project. At later stages decisions will also need to be made about how to analyse data, what it means in relation to the topic and one's practice, and how best to compare the sense which one makes of it with other evidence which might be available, and with what other people make of it, including others involved in the research. The implications for one's practice will be part of the 'making sense' stage of the research, and decisions about any changes in practice which may be appropriate will follow. So will decisions about how best to report the research to others for whom it might be significant and helpful in their practice.

However, doing research is not so neatly sequential as that. It is not uncommon for example for data which begins to emerge to cause a rethink on the way the problem or topic was defined. Or, evidence might arise which is so convincing that some immediate changes in practice can be justified or required, thus changing the situation and the focus. It is well worth recording decisions about these factors in the research process, perhaps in diary format, so that they can provide a systematic or disciplined framework for the project. If amendments to the procedures are made they would be made on justifiable bases, in response, say, to experience in implementing the research methods adopted, or preliminary analysis of data.

TOPICS AND FOCUS

Sometimes the most teasing of all steps in the research process can be the first one, deciding what topic to research. It can be teasing because one has only a vague sense of a problem, an uneasy feeling about events, or a sense of evidence which is hard to pin down. The topic of a research project might stem directly from the general purpose of research: a concern to monitor events with the intention to develop insights into one's own ideas, or classroom and school processes in order to improve insights, understanding and practice, but without a specific problem in mind. In case study which is not action research an equivalent to this would be a desire, perhaps, to understand social interactions or other phenomena in educational institutions. In policy analysis it might be to understand the origins and locus of curriculum ideas, especially where there is an apparent conflict of viewpoints and beliefs affecting school policy and practices.

In other circumstances there may be a clear imperative to carry out the research in a particular field; in the cases in this book, the introduction of the design and technology curriculum requirements. Sometimes it is easier to focus on aspects of the classroom other than one's own ideas and actions, but a main purpose of action research is to develop understanding of one's own educational principles and practice. That can be more disquieting, but also more rewarding, than looking outside oneself. It may mean opening up what one thought were certainties to question, examining what one has 'always known', as well as exploring doubts and dilemmas which were already inherently uncomfortable. There are many potential topics, out of which it is necessary to decide what is most relevant and worthwhile for developing understanding, as well as what is feasible (in terms of having access to data, for instance where young children are involved, and in terms of time and available resources).

Selecting the purpose and focus of a research project associated with classroom practice is a necessary but sometimes lengthy process. It is easy to imagine grand projects among the complex events of classroom and school life, less easy to select a focus which is manageable within a research topic. Topics for research are not independent of many related (and in reality inextricable) features of teaching and learning. Deciding on a topic may be helped by considering the questions: What, ideally, do I most want to know/investigate? Why is the topic significant for me? The task described below can focus attention on initial research ideas.

Task Write a one-page presentation of what you want to investigate, including the educational aims involved, and say why it has priority and significance. Try to make the topic clear, realizable (i.e. what you can do) and manageable (in the time available).

Recording these initial ideas for a research project and being prepared to review and amend them as one proceeds with the research is a worthwhile discipline for refining ideas, and for judging if a project is feasible. In the case of planned actions this will include wanting to know if events turn out as one intended them to, or to know what will happen as a consequence of (or despite) one's actions. Such research questions, hypotheses, problems, or issues take one's thoughts into the world of predictions, speculations, anticipations, and enquiry. This is a world of puzzlement and uncertainty, and needs fluidity in thinking and initially a willingness not to 'know' too much, but rather to want to know. However, we usually know some things based on firm evidence about a case, or about the situation, and have reasonably well informed judgements about other things. It is worth recording these at the initial stage of a project.

In fact, the topic of enquiry and the reason why it is significant often stem from informed judgements about what may happen or may be happening in practice, something one suspects is happening, and so on. From such impressions or aspirations a wish to know more about a situation or problem in practice arises, and provides the impetus for the research. Describing a situation/problem as it appears to be, or as one thinks it is, and its institutional or social context, provides a way to clarify ideas about the problem itself and one's awareness of evidence related to it. This can begin by considering the question: What is the situation/background to the topic? What do you think you know about it? Pursuing a task similar to the following one can be helpful.

Task Describe the situation, briefly and impressionistically, in writing. Provide whatever evidence is available. Consider how trustworthy the evidence is (is it convincing?). Record and file the description to provide a comparison against later evidence.

PRINCIPLES

Research can be threatening, because it involves questioning deeply held values, opening up one's practice to scrutiny, or changing one's perceptions of the world. Or, as in cases reported later, it means starting from scratch, accepting not knowing, and taking risks associated with learning new knowledge, testing new teaching strategies, and learning how to carry out classroom research. Action research almost invariably involves other people, their thoughts, ideas, perspectives, values and actions. It is important to follow principles in the conduct of the research which will ensure, so far as possible, that it is effective, worthwhile, and at the same time ethical. The principle of openness can be a sound basis for the conduct of observations, research discussions, interviews, or reports, to ensure a supportive climate and collaboration with those involved. This principle should guide the research process with colleagues and pupils, wherever possible in my view, i.e. one should be open about the subject, purposes and uses of the research with everyone concerned with it.

Listening to other people's ideas open-mindedly also appears to help focus upon one's own concerns with an open mind. The principle of open-mindedness to one's own misgivings, misapprehensions and misdemeanours is also a valuable asset to have. Sharing and scrutinizing ideas and evidence appears to be mutually beneficial, in the sense of opening up more information for the researcher, and helping others to think about their part in the matters being researched. But this involves making one's own ideas explicit and acknowledging that they may be based on very different values from those of others involved.

A process of constant negotiation and interaction is needed, and this can be rather destabilizing. It should be destabilizing, in the sense of self-conscious questioning of practice. But to be constructive, the conduct of research should also seek to enhance confidence, knowledge and skills within the focus of study, and enable participants to evaluate events in a supportive, sharing atmosphere. The following aims and practices can enhance openness and open-mindedness, if they are adopted as principles of procedure during a research project:

- identifying and sharing the aims of the research, one's own needs in carrying it out, and helping others to understand what the work is about and the problems it entails;
- listening to others' reports of classroom events and helping them to find ways of sharing them;
- asking questions about the research topic and others' work related to it in ways which will help to focus on it without threat to the self-confidence of others or oneself;
- applying constructively critical questions to one's own work, and sharing those thoughts with others.

A principle of anonymity may help to achieve openness by ensuring that, where it seems appropriate, names of people and places are changed to protect individuals from possible consequences of sensitive data being communicated to others. If the people involved know that they will be protected in this way, access to such data might be gained more easily. If young children are involved and negotiation of the use of data is not feasible, anonymity can provide a general safeguard. It is a safeguard principle, and in many instances will be quite unnecessary. The researcher must judge each research situation, and raise the question of the right of those involved to control the disclosure of real names.

The promise that information will be treated in confidence can also, sometimes, help one to gain evidence which would not otherwise be available. However, the principle of confidentiality may also compromise the principle of openness, and make it impossible to use data directly. On the other hand one may feel able to open up one's own work and ideas only in confidence, or in confidence to certain people. It is worth considering this principle in relation to research, but generally speaking it reflects, and represents, a rather negating stance towards the development of teaching within a colleagial setting.

Which ever of these principles is adopted will affect the negotiations which will need to be carried for the research to take place, if others are in any way involved. Declaring the principles on which one wants to proceed will ensure that individuals know what is happening, and what their rights are, including a right not to participate; to scrutinize any data they provide; to amend or correct what they say; or to restrict the circulation of evidence they provide. They will also ensure that the conduct of the research will be open to review as it proceeds, and that the outcome of the research, where possible, will be shared and open for discussion among interested people.

DATA

When first embarking on research the question 'What is data?' is often asked. Put simply, a sense datum is an immediate experience of the physical world – sound, colour, feel, movement, etc. More usually in the social research sense a datum is a single piece of information, sometimes heard, sometimes seen, or maybe written and read. Data are a series of such information about people's ideas, actions or interactions, which can take many forms. Policy documents, or documents produced in planning teaching, for example, contain information. Visual aids or worksheets often communicate instructions, and constitute data which is part of an interaction between teacher and pupil. Discussions between colleagues, or with pupils, and pupil interactions, represent a constant flow of data within the day to day activities of teaching and learning. Classroom layouts and resourcing convey important information about the physical construction of learning environments, as do the timing and phasing of teaching and learning projects. Teacher interventions in supporting pupil learning (for example, physical movement, eye contacts, or helping with pupils' work) are also data. Pupils' thoughts about their work or themselves in relation to it, and teachers' thoughts about pupils and their work are another source, as are written texts such as syllabuses, teachers' notes or pupils' exercises ... the list could be almost endless.

Often different kinds of data are most useful in combination to gain a more detailed view of events. Sometimes they are dismissed as not of interest, or not significant, so it is important to remain sensitive to their potential. However, social settings like classrooms are so full of data that it is also important to ensure that a particular research interest is allowed to steer attention towards the most significant data for that topic, and to guide its collection. The relationship between the topic and the question of what data is significant is one which will perhaps perpetually pose a problem in carrying out a research project. There will usually be far more data in a situation than one can handle; it has to be recorded urgently if it is not to be lost; and until it has been recorded and thought about it is often not evident whether it will be useful or not.

One solution to the dilemmas this can pose is constantly to check the data against the research topic and the aims or questions which are priorities within the topic. It is well worth establishing these aims and questions at an early stage, and in the case of action research ensuring that they relate directly to the teaching aims and their consequent practices which are subject to investigation.

In the first place, data may be evident in a situation, for example through observation of events in a classroom under normal conditions, or in students' work in response to tasks. On other occasions it might be less evident, and have to be solicited, as in the case of requesting an interview with a respondent to hear about why they act in a particular way. Data might even have to be engendered, in the sense that teaching situations can be devised or managed in order to generate new kinds of responses from students. That kind of circumstance would be common in testing new teaching materials or practices.

It may be necessary first to negotiate access to and the use of data among others involved in a research project. Sometimes that will be followed immediately by recording data – for example, asking if a colleague will agree to be interviewed, agreeing future use of the interview responses, and proceeding to tape the interview; or asking for a questionnaire to be completed, or if photographs can be taken, and establishing how they may be used later, and so on.

The different forms in which data occur are likely to need different ways of acquiring them and collecting them together, or recording them.

Some of these seem to be self-evident, as in the case of photographs, video film, or tape-recorded interview. But in each case alternative ways of describing events would be possible, such as descriptive observation notes in place of photographs or film, and interview notes in place of audio-tape. In all cases records need to be kept of locations, occasions, and respondents, so that the places, times and people can be identified and fitted into the larger picture of the research at the stage of collating and analysing the whole of the data. Simple methods such as a box file or ring binder system, self adhesive labels, or stick-on memo notes, can prove invaluable in the organization of data so that sections of it can be retrieved or referred to easily when they are needed.

Several techniques of data collection from the tradition of quantitative research are available. They include testing; surveys; questionnaire; experimental programmes; clinical interviews; systematic observation; analysis of documentary evidence. The traditions of qualitative research include the analysis of documentary evidence; the use of written accounts of events, or statements; interviews (structured by pre-set questions, or unstructured and free-flowing); observation of situations and actions; participation in and concurrent observation of activities and events; group discussion; reflective dialogue; autobiography.

It is worth noting a difference between methods which generate data, techniques which can be used for recording it, and systems for organizing it. These are sometimes confused because they overlap. For example, setting a task for students would generate data in the form of their responses; they might write or draw a representation of those responses; and that work would be a recorded set of data. Usually in action research qualitative methods are used, as they associate more readily with small-scale case-study research. That does not preclude the quantitative option, which may be appropriate for particular purposes even in case study.

Providing an impression of a situation/problem, and further considering the purpose and topic of the research, will reveal something of what is not known, as much as or more than what is. A description of the context is also valuable for others who might want to compare the research with their own case-study circumstances. Consider the question: What do I already know that is pertinent to the research?

Task Describe the proposed teaching/professional action which will take place, so far as it can be predicted, and/or describe the problem/research question which you want to pursue.

Task It is then necessary to consider the following questions: What data do I need? Where can it be found? Decide in detail the data you need to collect. Selection depends on the nature of the question and practicality of access.

Task Consider the question: What techniques should I use for collecting, recording, storing data? Decide in detail how to record and file the data.

RESOURCES

The resources needed to conduct classroom research will vary according to the project, except for the use of time. As always in teaching this will mean managing time differently to allow for research, rather than finding additional time. That will be so whether one works alone or with the help of pupils or colleagues. Consideration of time is important in terms of the total time-scale of the project, as well as the amount given day-to-day for negotiating the research, gathering data, organizing and analysing it, writing, and reading. This will be a main consideration in deciding what to research, since any project will need to be manageable and realizable within the time-scale available and within the day-to-day time of a busy teaching role.

Task Consider the question: What should be the timing and circumstances of the investigation? Work out a feasible time-scale and time-use strategy for the project.

In addition, the resources to consider include suitable means of recording and storing data which will help to provide a picture of events: notebook or diary; files to store records; tape-recorder and audio-tapes; still camera and film; video recorder and tapes; materials for pupils' work; a word processor or micro computer with a database facility, etc.

Task Consider the question: what would be the most appropriate material resources for the project? Draw up a material resources list.

Perhaps the most significant resource of all is the availability of people who can provide the data needed, and who are therefore crucial to the project. Deciding which individuals to invite to participate, and how many, is another judgement which can be made only in relation to the topic, timescale, and so on. The group and the choice of setting that are relevant to the area under investigation needs careful consideration.

Task Consider the question: who, ideally, should be involved in the research? Draw up a list of the people to be invited to participate.

READING

There is a risk in doing research with 'personal value' that it will be done without reference to other relevant research or scholarship. That can result from lack of time or difficulty of access to resources. It can stem from a view of the 'uniqueness' of situations. Where relevant literature exists it can be a valuable resource. However, when engaged in reading, as well as when making notes from literature, and when citing sources in research reports to colleagues, it is important to bear in mind the kinds of documents which are used and the reasons they are being considered. Mainly the reason will be to provide a broader research context from literature for the case study. That will ensure that the research is done with awareness of other similar and related research or scholarship, and in cognisance of how the case study fits in that broader picture. But the picture may be complicated by the kinds of materials used and the standpoints taken by the authors.

Some differences in the nature of published materials are obvious. For example, a newspaper editorial will probably represent the particular prejudices and opinions of an individual editor. An education act represents legislation approved by Parliament, which may have been driven by a single political party. Not all differences are so obvious. A government Green Paper may be similar in tone to a White Paper, but they have different status in the legislative process. A research report may appear similar to a scholarly book, but the former should be based on empirical evidence, whilst the latter might focus on the analysis of ideas. Even in research reports and scholarly books, but more so in journal articles, it is important to distinguish opinion from warranted assertions based on evidence or reasoned argument. Wherever possible it is important also to know the political or ideological perspective of the author: radical feminists and white liberal males, for example, may well differ in their prior assumptions about what they see in the world, and the way they analyse and report evidence.

Reports from agencies of government such as the Office of Standards in Education or the Schools Curriculum and Assessment Authority may appear to come from the same source as Department for Education policy documents or official reports of committees, but they do not. The source and status of, and the ways of constructing, information and documents related to the recent education acts (curriculum and assessment; teachers' pay and conditions; school organization; appraisal, etc.) is now very diverse. Legislation; statutory orders; consultation documents; consultation reports; administrative memoranda; DFEE Circulars; LEA and school policy documents; research reports; propaganda pamphlets; news reports; scholarly argument, and so on, need to be treated in their own right as forms of data emanating from particular human sources, and subject to scrutiny and analysis.

With this in mind, reading should be undertaken as an enquiry in itself, with the prospect of treating written material as evidence, as printed impressions of people's ideas, information and events. Reading does not provide a straightforward source of understanding, there for borrowing. It provides a range of sources, and alternative viewpoints, to which the reader will bring his or her own. The interface between the reader and those sources is one arena for intellectual struggle, with its attendant confusions, demands of information handling, and the coming to terms with new ways of thinking.

The extent of these sources is such that it will be necessary for each teacher researcher to construct his or her own list of materials encountered and used. Some might be provided, others suggested, still others referred to, during courses of tuition. Others will be uncovered during library sessions and field work. A systematic way of recording published written material from the beginning is recommended, such as a card index file (a supplementary system for news clippings and working documents may be used as part of data storage).

DESIGNING AND DOING THE RESEARCH

If the guidance so far has been followed, many of the decisions needed about the research design and strategy will have been made. At this stage it is worth recording an outline research proposal and inviting colleagues to consider it, to ensure nothing has been forgotten, or to have a second view of what is being proposed. This can be used as a personal checklist, and can act as a public account of the thinking behind the project. It also provides a useful exercise in the concise presentation of ideas.

Task Create an outline of the project using the following headings:

- Topic and focus.
- Why is it important?
- Underlying principles for the conduct of research.
- The location for carrying it out.
- People providing data.
- Data to be sought.
- Methods to be used for a) generating data; b) storing data.
- Relevant literature.

The plan can later be used to help to organize notes and provide sections of any written report. It may now be evident that the topic/problem is a fluid one, that views of it can change. That may continue as the gathering of data (which often holds surprises) gets under way.

Task Carry out the research plan and work out a strategy for continuing data gathering to complete the first phase of the research.

ANALYSING DATA

At this stage a project will simultaneously involve generating and collecting data, interpreting it, and organizing it. It is difficult in practice to separate these actions, but some initial analysis is likely to occur during the gathering and storing of data. A mass of data is not difficult to collect – in order to make any sort of contribution in an investigation it will have to be analysed and classified in some way. For example: by categorizing the data in relation to the questions posed; by looking at possible relationships between the data collected about teaching intentions and actions, and the data collected about what the students did in response.

Making sense of data can be based on a search for patterns in the responses of informants or the actions observed. These might well be subsets of information within the research questions being asked. To use a simple example, the question: 'What instructional strategies did I use?' might generate data on a range of categories, such as giving instructions, inviting ideas, setting written tasks, dictating notes, asking questions. Within the category 'asking questions' there might be subsections depicting types of questions: those requiring factual knowledge; those setting problems; those inviting opinions, and so on. As patterns, issues and sets of data begin to be recognized they are translated from tape-recordings, pupils' work, notes, transcripts, documents, and such, into evidence.

Interpreting data also requires its systematic ordering, but this process of analysis can be devised as a simple discipline which begins with the labelling of documents, transcripts, etc. (name of person interviewed or observed, date, place of event). Using card dividers, plastic wallets, and numbering pages, paragraphs and any other obvious sub-unit of a file or document can also begin as data is gathered. These can be indexed, or colour coded, so that if one wants to locate evidence of (to use the earlier example) particular kinds of questioning, or gender issues, or particular attitudes, etc. it is possible to do so systematically.

Task Organize the data into manageable, accessible, and 'presentable' form.

Now is the time when sorting out the data will lead to a process of selecting from it. This is a time of making more sense of those first impressions, of confirming the evidence first uncovered, by adding to it from the data.

Distortions can also occur in this process, through the selection of data which one 'wants to find'. At this stage it is possible to begin to summarize one's ideas which result from thinking about the evidence. Consider the questions: To what extent do your ideas match what you thought might happen or was happening before the investigation? Has something come to light of which you were unaware? Have you been alerted to any assumptions which were being made, say, about teaching, or about how the students might respond?

This will mean deciding what to leave 'on file'; whether some might form appendices in a report; and which to include as most significant when reporting. It will also mean deciding appropriate 'formats' – tables, charts, extracts of interview transcripts or field-notes, photographs, etc. Decisions about selection and format will be influenced by interpretations, and they will also influence the way the presented data is discussed. Preparation of a draft report based on the data, but adopting a critical perspective on what has been done so far, is a worthwhile step towards making the research available to others.

Task List the main categories of evidence which the data provide.

TAKING ACTION

The major purpose of the research is likely to be to inform one's own understanding, to help to develop practice on the basis of evidence which is more substantial than the usual impressions of classroom life, and to take better-informed action. Changes in the detail of existing ways of doing things can result from the initial analysis of data. Replacing existing ways of doing things with different ways in a wholesale manner is also possible. Within an innovation, though, the process is about testing new ways of doing things, and problems arising in the execution of a teaching plan can also be solved as evidence begins to accumulate. Proposals for action which are based on the research evidence can include statements of the aims of the changes or adjustments, and the reasons for them. These would be in the form of implications derived from the evidence, or recommendations for continuing and reinforcing particular actions, or for implementing changes. Like all proposals for professional action, these will take account of the evidence in the context of the situation, and will include immediate, short-term, and future long-term possible actions. Action plans may form the final part of a research report to colleagues. (They could include further proposed research if it is felt that more evidence is needed.) They will be based on the construction of sound argument derived from the research evidence, and justified on that basis.

A RESEARCH REPORT?

Creating a research report provides an opportunity for a critical review of the action, the research, and proposed actions stemming from it. It enables further consideration of some issues in the conduct of the research itself, as a means of enhancing the skills of action research. Getting a report into shape will depend upon its purpose and potential audience. In the case of reports submitted for advanced award bearing courses some formal frameworks may be in place, fitting the conventions needed for assessment purposes. Some guidance on the possible format and presentation of research reports, which will help others to read the case study in context, follows.

- The report should be typed or written on A4 paper. Pages should be numbered.
- A title page should indicate: name and school; a chosen title; the date.
- The second page should contain a single paragraph summary of what the report is about.
- A contents page should indicate the sections of the report, references and supporting materials. The use of sections, headings and sub-headings is often helpful in the report itself.
- Conventional structures of research reports often include:
    Introduction to the problem/issue
    Outline of the written report: a 'route-map' for the reader
    The issue in the context of literature/theory
    Methodological procedures (and the case or sample)
    Description of events/teaching/case study
    Interpretation of observations/analysis of data
    Theoretical considerations
    Summary/applicability and practical implications.
- Within the report where reference is made to supporting materials the sections in which these can be found should be clearly identified. Where certain types of materials have been gathered, such as pupils' writing, drawing, photographs, records, etc. Illustrative examples may be used to good effect within the report itself.
- Where reference is made in the text to a book or article it should be identified in the form: (Shepherd 1995) within the text, and listed at the end. If direct quotation is used it should usually be indented, and author and page number given in the text (Shepherd 1995: 19).
- References should correspond to a list at the end of the report, in alphabetical order by surname of authors, and using the following conventions: (for books) Shepherd, G. (1993) *Do It My Way*, London: Cut Price Press; (for chapters in edited books): Shepherd, G. (1993) 'Do it my way', in J. Major (ed.) *Managing Ministers*, London: Cut Price Press; (for articles in journals): Shepherd, G. (1991) 'Do it my way', *British Journal of Educational Change* 99 (1): 6–9.

- Supporting materials should be carefully organized and presented as appendices to allow easy access. This will obviously vary according to the nature of the material. It is advisable to be selective in the amounts included, to avoid repetition, and to ensure that only material which supports the report itself is used. These need to be cross-referenced in the text, in order that the way they are used to support arguments in the report can be seen easily.

It may be helpful to carry out a self-review of the report. Consider the following questions:

Was the 'topic'/problem adequately defined, or did it change?
Were the methods adequately employed to provide trustworthy data?
Are you aware of how elusive evidence was/is?
What have been the methods and effects of selection and organization of data for reporting?
Is there any indication that you have relied on belief, assertion, or unsubstantiated claim in the report?
How have you extended the critical analysis of the data, in order to elaborate the picture/account as you now see it?

Task Prepare a written report based on the work done so far, and guided by the notes on layout and organization and on the analysis of data.

QUALITY IN ACTION RESEARCH

Judgement about the quality of research can be made in direct relation to its value for improving teaching, or for developing one's understanding of a situation. Judgements are often based on a written report, particularly for advanced award bearing courses. In those circumstances the research needs simultaneously to meet academic criteria and to have practical relevance to work in the classroom and school. I believe those needs will both be met by the conduct of research which is rigorously conducted, relevant to the practices of the participants, and reflexive in terms of open self-criticism of the ideas behind it. Such research can be supported in group activities among colleagues in schools. Some criteria can be used as a guide to judging quality. For example, consider whether the work provides evidence of:

- the researcher having identified a theme, issue, question
- the researcher having developed exposition, argument and ideas
- discussion, analysis and reflection, which goes beyond description, unsubstantiated opinion and/or rhetoric
- the grounds on which sense has been made and understanding gained from the analysis of evidence and/or experience

- awareness of the conceptual and theoretical background and literature relevant to the topic

These five criteria came from an advanced award bearing programme in higher education and are intended for use in judging written reports. Distinguishing between quality during and within the research processes, and quality within a research report is difficult. It is important to be alert to this distinction. Going beyond the quality expected of a research report, and seeking to define quality in the whole of a research project, I invited some teacher researchers I worked with to suggest what credible and creditable teacher research would need to include to meet their quality approval. The following list of criteria was created and subsequently published (Tickle 1995):

- credibility, established by the voice of the researcher being made public
- inclusion of the researcher's values, beliefs and assumptions
- clarity of the research question or issue, the purpose of the research, and its process
- demonstration of the importance of the research as justification for doing it
- addressing an issue which is of interest to others in the educational community
- assurance that practical action strives to achieve educational aims
- making of explicit connections between the research and the learners to whom educational aims were directed
- incorporation of revisions in the direction of the research and changes in practice resulting from it
- demonstration of a self-critical stance towards practice and research
- bringing multiple perspectives to the data, and ensuring accuracy in its handling through self-checking
- presentation of sufficient and convincing evidence to support assertions and claims made in research reports or in practice
- conclusions following directly from evidence
- transferability to other situations – something similar should be capable of being done by others
- inclusion of new questions which arise from the research
- characterizing the work as unfinished, a continuing venture
- provocation through the setting of challenges to others and oneself in the ideas presented

These do not define the internal standards of each quality criterion, i.e. does not declare that 'complete assurance'; 'total clarity'; 'fully inclusive'; 'masterly demonstration', of the appropriate rules of rigour is required. That is also true of criteria gleaned from the international literature on educational action research, which I have used to compare with that other list (Tickle 1995):

- *prudence*   practical wisdom and the capacity to judge the most profitable courses of action
- *openness*   records, representations of action, reports of underlying values
- *open-mindedness*   reviews of evidence of these, responsiveness to reviews, reinterpretations of accounts
- *communal self-reflection*   exposure of prejudices, both as practitioner and as researcher; engaging in collective examination of prejudices.
- *courage*   in exposing curriculum proposals and practice; and in exposing research endeavours
- *growth*   a willingness to change and acknowledge change
- *contemplation*   living with the fermentation of understanding, especially of the theory–practice relationship, while seeking maturation
- *overview*   appreciating the interdependence of phenomena
- *configuration*   making meaningful form from complex information

Put this way the criteria are rather abstract, but their essence is of the reflective practitioner, determined to do the best possible for students, seeking continuous self-improvement, and working for the communal improvement of education. The contributors to this book were carrying out research as a basis for teaching quality. The full extent of their work, for reasons of space, has been subject to editing. Some more than others have been refined to summaries of what they did, and how they did it. Behind all of them lies a lot more data, and a great deal of reflectiveness about the research methods and processes.

FURTHER READING

Altrichter, H. (1986) 'Visiting two worlds: an excursion into the methodological jungle including an optional evening's entertainment at the Rigour Club', *Cambridge Journal of Education* 16(2): 131–43.

Altrichter, H. (1993) 'The concept of quality in action research: giving practitioners a voice in educational research', in M. Schratz (ed.) *Qualitative Voices in Educational Research, Social Research and Educational Studies*, London: Falmer Press.

Altrichter, H., Posch, P. and Somekh, B. (1993) *Teachers Investigate Their Work*, London: Routledge.

Bell, J. (1989) *Doing Your Research Project*, Milton Keynes: Open University Press.

Carr, W. and Kemmis, S. (1986) *Becoming Critical: Knowing Through Action Research*, Lewes: Falmer Press.

Clarke, J., Dudley, P., Edwards, A., Rowland, S., Ryan, C., and Winter, R. (1993) 'Ways of presenting and critiquing action research reports', *International Journal of Educational Action Research* 1(3): 490–91.

Elliott, J. (1980) 'Implications of classroom research for professional development', in E. Hoyle and J. Megarry (eds) *Professional Development of Teachers*, World Year Book of Education, New York: Kogan Page.

Elliott, J. (1985) 'Educational action research', in J. Nisbet (ed.) *Research, Policy and Practice*, World Year Book of Education, New York: Kogan Page.

Elliott, J. (1988) 'Educational research and outsider–insider relations', *Qualitative Studies in Education* 1(2): 155–66.

Elliott, J. (1989) 'Educational theory and the professional learning of teachers', *Cambridge Journal of Education* 19(1): 81–101.

Elliott, J. (1991) *Action Research for Educational Change*, Milton Keynes: Open University Press.

Hopkins, D. (1985) *A Teacher's Guide to Classroom Research*, Milton Keynes: Open University Press.

Kemmis, S. and McTaggart, R. (1982) *The Action Research Planner*, Victoria: Deakin University Press.

McKernan, J. (1988) 'The countenance of curriculum action research: traditional, collaborative, and critical-emancipatory conceptions, *Journal of Curriculum and Supervision* 3(3): 173–200.

McKernan, J. (1991) *Curriculum Action Research*, London: Kogan Page.

Phillips, D. C. (1987) *Philosophy, Science, and Social Enquiry*, London: Pergamon.

Rudduck, J. and Hopkins, D. (1985) *Research as a Basis for Teaching*, London: Heinemann.

Schon, D. (1971) *Beyond The Stable State*, San Francisco: Jossey Bass.

Schon, D. (1983) *The Reflective Practitioner*, New York: Basic Books.

Schon, D. (1987) *Educating the Reflective Practitioner*, London: Jossey Bass.

Schostack, J. (1989) *Qualitative Research: A Guide to its Principles and Practice*, Norwich: School of Education, University of East Anglia.

Schratz, M. and Walker, R. (1995) *Research as Social Change*, London: Routledge.

Somekh, B. (1995) 'The contribution of action research to development in social endeavours: a position paper on action research methodology', *British Educational Research Journal* 21(3): 339–55.

Stenhouse, L. (1975) *An Introduction to Curriculum Research and Development*, London: Heinemann.

Tickle, L. (1989) 'New teachers and the development of professionalism, in M. L. Holly and C. S. McLaughlin (eds) *Perspectives on Teacher Professional Development*, Lewes: Falmer Press.

Tickle, L. (1994) *The Induction of New Teachers: Developing Reflective Professional Practice*, London: Cassell.

Tickle, L. (1995) 'Testing for quality in educational action research: a terrifying taxonomy?', *International Journal of Educational Action Research* 3(2): 235–38.

Winter, R. (1989) *Learning from Experience: Principles and Practice in Action Research*, Lewes: Falmer Press.

Chapter 3

Combining design, technology and science?

Martin Bayliss

Our curriculum structure had been partly 'secondary based' with some specialist teaching, particularly in the upper years. This was being reviewed because of loss of teaching staff. My responsibility was for coordinating the introduction of design and technology, which had never been on the timetable, and leading all the staff towards teaching it. Early efforts included consulting other schools in the locality and joining a local coordinators' group (made up mostly of technology 'specialists'). After careful consideration, school policy defined technology as 'cross-curricular', to be taught on a thematic or project basis, to enable the programmes of study to 'fit in' with other subjects.

Staff in-service was devised around a design and technology 'Orders Treasure Hunt', seeking to match the new demands to project work and other subject programmes of study. A central resource was established, with access for all teachers. A positive response from staff seemed evident. However, there was no detectable increase in the use of the resources, and little evidence of technology taking place. I decided to try to lead by example, reorganizing my art and craft sessions to include new making skills: Jinks frames, Lego technic, and graphic design, alongside the usual painting, collage, and fabric work. I attempted simple design and make activities with my class and displayed the results as an encouragement to other teachers.

Two technology projects were developed with my class: a computer controlled bridge, and a novelty lighting tie. They both involved design and technological experiences which helped to develop knowledge and understanding, as well as introducing new and challenging making skills. The pupils were very enthusiastic and I judged their results to be of a high quality, both in what they learnt and what they produced. Both projects generated a lot of interest from within and beyond the school: the first particularly from the local Port Authority; the second from a marketing company.

THE BRIDGE PROJECT

There are two lifting bridges in Great Yarmouth, both controlled by computers. A computer interface was bought by the school and I was asked to build a model to demonstrate its potential. The idea of building a lifting bridge was well received by my class. The Port Authority was very helpful, and supplied a set of plans and copy of the specification for one of the real bridges. The local library has a file of press cuttings of its construction. Our model took six weeks to build. Pupils were divided into work teams and set specific tasks. Two teams worked on the plans; they decided a scale of 100th and drew a plan and side view of the bridge. Other teams made small sections of wood and card, carefully guided in this their first wood construction work. Lifting gear for the barriers and main bridge section were the biggest problem. Several attempted gearing design using Lego Technic, but there were problems with gears meshing under stress. A windscreen wiper motor, with its own gears, was eventually used to lift the carriageway.

All the making and painting was done by the children, and every child in the class was involved. Four worked on the control programme on our BBC computer and after some teething problems with loose wiring, the bridge actually worked. The children had experience of construction techniques, design decisions and computer control. They also had an insight into the workings of an important local landmark and its sophisticated modern technology. The finished model attracted a lot of interest from other classes, parents, and teachers from other schools. The local newspaper featured an article, and the Port Authority quarterly magazine included photographs and details of our work. There was some speculation as to which of the bridges worked more efficiently – ours or the real one. As for ownership of the project, all children had contributed, so all shared in the success. Some pupils were overheard referring to the full size version as 'their bridge'.

THE CHRISTMAS TIE

An in-service course asked us to design and make an item for personal adornment for a festive occasion. I decided on a Christmas tie and sketched a range of designs, to squirt water, spin round, or light up. I consulted a costume hire shop, a professional clown, and an electronics engineer. The chosen design was a clip-on tie with a felt Christmas tree, and two simple electronic circuits sewn in: one controlling a series of flashing light emitting diodes (LEDs), the other playing seasonal tunes. I was very pleased with this first attempt at designing with electronics.

The tie was great fun and the idea was shared with my class. Electrical components were bought and pupils spent lunchtimes soldering their circuits. They thought up a range of uses for the flashing light circuits, resulting in flashing bow-ties, headbands, brooches, badges, table decorations, and even a flashing baseball cap. The artefacts were worn in school at Christmas events and attracted a lot of interest. There were suggestions that we should go into production to raise funds.

An end-of-term school review of technology teaching revealed that I still had a lot to learn about promoting the subject across the school. Comments such as 'that's very good, but I couldn't do anything like that', and 'I wouldn't know how to begin, I just haven't got the skills', and 'I just haven't had the time, we have enough trouble with the other subjects constantly changing', were offered to me quite openly. The comment about shortage of time was supported by many colleagues, one of whom had trialed model making, but summed up her experience with: 'the making part takes so long, I had to send the models home unfinished, otherwise I wouldn't have done anything else.'

As coordinator, a lead seemed essential. Perhaps the concern about time was justified. I had devoted a disproportionate amount of time to the two projects, maybe at the expense of other subjects. They had taken up lunchtimes and after school hours. They did not exemplify a cross-curricular approach within classroom time, but rather an extra-curricular option for the enthusiast. They were also ambitious in their use of 'technology'. My Year 5 class were soldering electrical circuits with transistors and resistors. Although simple as such things go, these were far removed from the clip-on circuit boards usually used in school electricity projects. Despite the apparent success of my two projects they provided no evidence with which to argue against the opinions expressed by colleagues. I needed to demonstrate that design and technology could be taught in an integrated way, within the timetable, and without making unreasonable demands of the teachers in terms of technical knowledge and skill.

The teaching topics for all subjects are prescribed by curriculum coordinators. Many of the Spring term topics were either undergoing revision or were completely new, because of the cross-curricular policy. A history topic of exploration and encounter could have provided a good opportunity for design and make activity, but I was asked to link history with a new computer simulation programme about world exploration. Design and technology wasn't the only new thread we were trying to weave into the curriculum. The integration of information technology within the short history lessons presented enough complications without adding a design activity.

A science topic on water was chosen for my attempt to demonstrate the potential of integrating design and technology with other subjects. There were some practical reasons for this choice. Science is the longest single lesson on my class timetable – a whole afternoon. We have use of the whole of the 'shared' activity area as well as our own small classroom for this session. There is easy access to both science and technology stores. And I was already familiar with the content and activities of the water topic.

It was based on Nuffield Science cards, simple experiments, set out in clear language with step-by-step instructions. Practical lessons use a laboratory atmosphere, using correct scientific equipment. The card titles for Year 5 were: *Keeping ice cool*; *Melting at room temperature*; *Small icebergs*; *Salt water and fresh water*; *Floating in different liquids*; *Floaters and sinkers*; *The Cartesian diver.*

I decided on a two-phase project based on the *Titanic* disaster. The first phase asked simple questions about the disaster, leading children into discussion and scientific investigation. Each session ended with an evaluation of the methods used and the answers given:

What is ice and where does it come from?
Why is ice so dangerous to ships?
What was the *Titanic* made from and why was this unusual?
Hundreds of passengers drowned. What could have saved them?
How does the salt in water affect the way things float?

Phase two was a five-week design and make task in which pupils could apply their new knowledge, link making activities with the context of the topic, and engage in evaluation of their design. They were asked to design and make something which could tell or help to tell the story of the *Titanic*. This might be a model of the ship or of a mini-submarine which was used to rediscover the wreckage. It might be a story book, a collage, or a wall display. They were also asked to make a presentation of work to the class and answer questions from peers.

The research perspective I chose was one of evaluation of the teaching project, with an intention to report to colleagues. At the time I hoped to demonstrate to them what I anticipated would be the success of integration of the science and technology. The data I sought and methods I used for recording evidence included: a diary of classroom activities and observations; pupils' written reports at the end of each activity (a normal part of science lessons); video recording of children working and of final presentations; and interviews with groups after activities.

The first few weeks raised many questions and the pupils' responses led to good scientific investigations – too many to report here. By week four pupils were asked to design a device which would help victims of a sea disaster to survive. The survival aids would have to pack away in a small space, be light, cheap, and easy to use. A range of materials was provided, and pupils asked to make up tests to find the most suitable materials, and to present their ideas as a labelled diagram. An example from my notes of Katie, Sarah and Michelle's design, illustrates what followed.

The girls tested and used a variety of materials which would float and provide warmth. They had considered the need for waterproof joining, had used their knowledge of materials, and selected appropriate hand-tools. They were able to evaluate their work, describing and discussing their decisions and improvisations. They had not considered the size of their 'suit' or how it would be put on.

These activities indicated some overlap between scientific investigation and the design process. Knowledge gained through investigating was applied to a practical problem. However, there were seven groups of children, and each attempted at least five investigations. Apart from the three girls, none had time to make up the full versions of the survival aid designs. This first phase of the project had represented a different approach to my science teaching but did not demonstate my aim of integrating the subjects within the time and resources available.

The phase two design and make task brought a wide range of proposals: twenty model boats, twelve story books, two submarine models, a play, puppet show, collage, wall display, and a presentation in school assembly. However, the ideas were not linked to the scientific investigations which had been undertaken, were crudely presented, and in my view quite inadequate as design proposals. There was no evidence of research to discover or confirm details; some ship models had the wrong number of funnels; and there was no indication of the materials, sizes, scales, or colours to be used. For example:

MB   What colour will you paint the funnels?
PUPIL   Red I think, I'm not sure. Yes, red.
MB   Daniel, do you agree?
DANIEL   Yes, they should be red.
MB   Have you checked in the books?
BOTH   No.

Two weeks later, when their funnels were already painted red, these two very bright children discovered a picture of the *Titanic* with yellow funnels:

DANIEL   Mr Bayliss, the funnels are yellow. How can we paint over the red with yellow? It won't work.

Yellow sticky tape was the solution they devised, followed by more considered application of colour: red for the Plimsoll line, white for the superstructure, and brown for the decks.

Jodie, Charmain, Helen and Laura were impressed by a picture showing a cross-section of the *Titanic*. Translating this into a model was obviously too ambitious a project. They spent time discussing what to select that was manageable.

The range of ideas seemed fine, but I soon became aware that there was very little science in this. I asked them to remember the investigations, and to 'think science' in their designs. The first practical making session confirmed my fears. I kept asking myself 'where is the science in this?' as children were glueing Jinks frames and colouring book covers. I couldn't see how the activities related to phase one of the project. Helen showed me the lampshades she had made for her cabin. I commented: 'Very good. Did the *Titanic* have electric or gas?'. 'At last, science', I thought. It wasn't related to the investigations but I was desperate to make the links.

The project began to go 'wrong' from a very early stage. The Year 5 staff planning meeting for Spring term science had been postponed from the Autumn, and took place three days into term. I presented my proposals for the project and outlined the ways in which science and technology could be combined through the *Titanic* story. My colleagues seemed unwilling to give up the approach with which they were already familiar. This was summed up by one colleague: 'With the pressures we are under at present I have to prioritize. Science is not a high priority of mine at present, and I haven't even begun to think about technology.' Similar comments came from other experienced colleagues, and represent something of the overload of change they were having to face.

The science coordinator expressed her support for the project, then continued the meeting with a list of additional content which would need to be covered in order to fulfil the requirements of national curriculum science. The addition of so many topics, at this late stage, dealt a blow to my morale. How could one plan ahead? There were twelve science lessons in the term, and we were being asked to cover fifteen topics. The content specified would not be covered by my project. I had decided to extend phase one, include additional material, and steer investigations towards the additional knowledge content.

In the event, as I have shown, we needed a different design activity if that was to be achieved. Earlier ideas had been to design and make an unsinkable boat; a boat to carry a 1kg cargo; an underwater lift which works; or a device for lifting the *Titanic* from the sea bed. I had actually set a task about telling the story of the disaster. I couldn't help wondering if my other ideas might have led to a scientific line of activity. As it was I had failed in my aim to demonstrate that science and technology could be taught simultaneously. A large investment of extra time and resources was needed. My wondering led me to reflect on some of the arguments I had read about the organization and structuring of the school curriculum. What is valued and taught, and how we teach it, are bound up in the discussion about the nature of knowledge itself.

There are different perspectives of 'knowledge', which might lead to different conclusions about curriculum planning. My teaching/research project began from within a subject-based, content-led curriculum. Syllabuses prescribed the content, sometimes on a weekly basis and fixed programme: week three, Spring term, all Year 5 pupils worked from Nuffield Science 'Small icebergs'. National curriculum subjects mainly fitted the school's former pattern of describing distinct content detail for each traditional subject. The inclusion of design and technology as 'cross-curricular' was somewhat at odds with this. The rest fitted Hirst's (1974) 'rationalist' view of the curriculum, with distinct subject content and ways of investigating, experimenting, thinking and expressing defined within clear boundaries. These are the 'disciplines' or forms of knowledge, each with their own unique content and structure. (He proposed mathematics, science, morals, religion, human sciences, history and aesthetics.) Perhaps this was the problem – science has strong boundaries. The identity of design and technology is more akin to Hirst's combined 'fields' of knowledge, such as medicine, engineering, or geography, which draw upon contributory disciplines, for the purposes of more practical application. It is less firmly held in our culture, and hadn't even a toehold in some parts of the school.

I was also conscious of other philosophical arguments about learning: ideas about 'empirical' knowledge which is constructed by individuals on the basis of their experience of the world; and the ideas of educational 'pragmatists' whose concerns are with how experiences can be made available to learners in the most effective forms (see Kelly 1986; Pring 1976).

Establishing design and technology as a discrete subject on the school timetable might be possible through the displacement of time allocated to other subjects. However, if the cross-curricular principle is upheld, and looser boundaries in subject content permitted, a different kind of curriculum review would be needed. A balance of learning 'experiences' which would serve the imaginations, explorations, and developmental needs of each pupil would need to be accommodated. The design process could be instrumental in redressing the balance of pupils' experiences and opportunities for learning, away from the content-based, teacher directed pedagogy. However, this would clearly be a sensitive matter. It would require a radical curriculum review, involving all members of staff in reconsideration of the ways in which the national curriculum has been designed and is being interpreted and implemented.

At the beginning of this project I was confident of achieving my aim of demonstrating that science and design/technology could be integrated. I was not 'investigating the possibility', or 'researching the problems' of a partnership of the two subjects. Failure was not a consideration, and I was confident of background support for my views: 'Some scientific principles might be more easily understood through a problem involving making something embodying or demonstrating the principle, or through children being allowed to design their own experiments' (Design Council 1987: para. 4.9). What the 'failure' has achieved is to raise issues, to promote discussion in school which goes beyond worries about inadequate skills, training or resources, beyond grumbles about subject content timetabling. These problems still exist and have to be faced. But we are now beginning to consider a more fundamental question of the nature of the educational enterprise, and the place of children within it.

FURTHER READING

Assessment of Performance Unit (1987) *Design and Technological Activity*, London: HMSO.
Barlex, D. (1991) 'Using science in design and technology', *Design and Technology Teaching* 24(1):148.
Design Council (1987) *Design and Primary Education*, London: The Design Council.
Hirst, P. (1974) *Knowledge and the Curriculum*, London: Routledge & Kegan Paul.
Kelly, A.V. (1986) *Knowledge and Curriculum Planning*, London: Harper & Row.
Pring, R. (1976) *Knowledge and Schooling*, London: Open Books.
Rudduck, J. (1991) *Innovation and Change*, Milton Keynes: Open University Press.
Thistlewood, D. (1990) *Issues in Design Education*, Harlow: Longman.

Chapter 4

Effective completion of technological tasks

Sue Cooke

The mixed age junior class of twelve girls and eleven boys is organized in year groups with nine in Year 4 and seven in each of Years 5 and 6. Most areas of the curriculum are undertaken by children working in year groups, although for both science and design/technology groupwork often presents the opportunity of working with children of other ages. All children study the same area of the curriculum at the same time.

The classroom is a 'Horsa' hut which measures 7m × 7m and once provided only dining facilities. There are few storage spaces available and only a restricted working space, so some equipment and materials have to be stored in the main school building. The room is used to serve hot lunches, therefore tables need to be cleared at 11.55 a.m. every day. There is a storage problem for models which are in the process of being made.

Children have access to a toolboard which provides six junior hacksaws, a hammer, two hand drills and drill bits, two steel rulers, six G clamps, six pairs of scissors. An A3 cutting mat, six safety knives, a Stanley knife, a large pair of scissors and a hot glue gun are also available in the room.

Expendable resources are very meagre: a few sheets of thick card, a few scraps of board, some dowelling and 10mm square section wood. Wood glue, UHU glue, Pritt Sticks and PVA glue provide the range of adhesives. All wood is stored in the main school building. Paint is mostly ready mixed liquid acrylic with some tins of tempera powder paint. Batteries, lights and motors are minimal. There are four incomplete boxes of Technic Lego.

Technology is undertaken one afternoon per week. A whole afternoon is preferred to single hour sessions because of the problem of collecting resources from, and returning them to, the main building at the beginning and end of each session. The class had had limited technology and design experience. The majority have a little experience of using woodworking tools, particularly Year 6 children.

As a newly appointed headteacher at the time of this research I was still learning how to combine the role of full time class teacher with the necessary management tasks that have to be undertaken. Previous teaching experience was in a middle (deemed primary) school. I have also worked as an advisory teacher in Information Technology. I was in a learning situation, both with a mixed age class and the need to teach design and technology, when this research was done.

As researcher and headteacher I was in a position to investigate problems and, as a result of the findings, make immediate changes to my own work and recommendations for change which may affect the other teacher's working practice. Also as I highlighted deficiencies in resources, I was in a position to make good the deficiencies within the budget limitations of the school. In this instance there was a particular problem which I wanted to address, which was: Why are the Year 6 children in this school not completing technological tasks as effectively as I would wish? At the time I described the source of the 'problem' as follows:

> The pupils at present do not work as effectively as I would wish. Many fail to finish the set task resulting in numerous unfinished models at the end of a project. The results often do not match my ideas of the original intention, and models often bear no resemblance to what the children set out to make. They fail to pay sufficient attention to finished appearance, and models look 'messy' so that even well built structures look unpresentable.

I had speculated on some possible reasons for the problem, construing that it could be caused by:

- misconceived classroom organization (lack of work space, storage of materials, etc.)
- wrong or lack of classroom resources (equipment, range of materials, etc.)
- mismatch of teaching strategy (groupwork, pairing, time devoted to discussing progress, etc.)
- misunderstanding by pupils of what is required
- lack of teacher/pupil communication
- unrealistic tasks
- lack of pupils' necessary planning experience
- undeveloped or lack of necessary 'making, skills
- immature appreciation of time available

The list itself suggests the nature and extent of my puzzle. I would need evidence before I could make a judgement on any one of these, but there were other possible reasons, which I recorded as follows:

> The desire to 'make' at all costs appears to be viewed by the children as the most important part of the whole process with little attention being paid to design and planning. Formal design work has not previously been undertaken before construction. The children have been

used to only a group 'discussion' in which the teacher introduced the idea of the work to be done. Tasks previously were carefully prescribed and supervised, thus making decision taking (which I was trying to foster) a 'new' experience. Limited use of 'hands on experience' over the last two years had also seemed to fuel the children's desire to hurriedly 'make things'.

Lack of planning and evaluation cause considerable concern. The children appeared to dislike any form of graphic preparation, either drawing or script. They didn't seem to perceive the connection between the design process and making process. All design work appeared to be viewed as a 'waste' of valuable 'making' time. Previous work had never involved any formal evaluation of finished products. Thus the children were not used to considering critically their efforts and without plans and designs they seemed to accept their results because they didn't have these with which to compare the final product. Where products didn't match original intentions there was no systematic way of judging the reasons, or evaluating their thinking, since the ideas were (seemingly) loosely formed.

The lack of awareness of time, in particular in relation to planning needs, was a problem highlighted. The amount of time allowed to complete a task did not seem to mean anything to the children. Projects which were far too complicated and which would obviously require more time than was available were embarked upon. The children then seemed genuinely surprised that the work was unfinished, but at the same time also seemed content to accept the situation.

Certain poor work appears to have been caused by the children's lack of, or under-developed, skills in using tools, particularly 'cutting' tools. Lack of knowledge of materials coupled with a preference to use wood regardless of its appropriateness seems to aggravate the situation. This seemed to link with the limited use of tools over the past two years and also some preconceived idea on the children's part that technology meant 'woodworking'.

Having identified the question and some impressionistic reasons for the problem I chose to focus on certain aspects of those reasons, and to design a research plan which would enable these aspects of the problem to be explored. This amounted to raising the most 'plausible' causes of the problem to the top of the list for investigative purposes. In doing so I judged that it was important when planning which data to collect to be aware of my own preconceived ideas: of what I believed classroom practice should be achieving, and why I believed this not to be happening.

The problem is not simply confined to one particular age group, but limiting the research to just Year 6 made the task more manageable. Since the enquiry depended on contributions from the children I also felt that the Year 6 children would be able to offer more constructive assistance than the other two year groups. Year 6 children were approached with a view to taking part in the investigation. It was explained what would be involved and all the children were willing to help.

All work undertaken by the children centred on the term's project of 'communication'. The specific Year 6 task for the term was to make an alarm system for the school, but prior to starting that brief all the children in the class undertook a number of activities which investigated various aspects of communication. The aims of the project and specific tasks which I built into it were to:

- incorporate combined experience and knowledge in a group work exercise;
- look at the importance of shape and colour in effective communication;
- allow pupils to discover how communication can be achieved through a variety of means;
- examine various ways of communication between people who may be separated by distance;
- encourage pupils to appreciate the most appropriate method of communication in a given situation; and
- examine the use of buzzers, bulbs and switches in the making of a simple circuit.

The children worked in pairs, threes or fours. Short projects were designed to give the children experience of using certain tools, materials and components. These included batteries, wire, bulbs, buzzers, wood, cardboard, paper, scissors, string, glue, cotton, wool, wire, paint, felt pens, coloured crayons, yoghurt pots, cups, and tubs; software – Pendown, Folio, Control S; hardware – Control Box, BBC and A3000 computers. Skill instruction was followed by the children performing a task which required the use of the 'new' tools.

The plan involved collecting data of the working practice and design products with the children throughout the Spring term. Classroom layout and observations of working practice, photographs of finished products, together with tape recordings of interviews were used as 'substantiating' evidence for cross-reference. Collection of data and observations were undertaken during the normal class technology lessons and some of the limited administration time I was allowed as headteacher. A series of questions were drawn up which focused on four specific areas:

- manufacturing (whether the children realized they were lacking in 'making' skills);
- time (whether the children's appreciation and/or management of time limited their capacity to plan a project successfully);
- planning (whether the children viewed their plans as an essential part of their technology work);

- evaluation (whether the children were 'happy' with their end products and on what basis did they make that decision).

The questions were designed for individual interviews with the children following the completion of a task done within the early stages of the topic. Some of the same questions were used following the completion of the term's task on 'designing an alarm'. Manufacture, time, planning and evaluation were chosen for several reasons:

- they were the four areas of the technology curriculum where I had definite views relating to the quality of the children's work and which I felt therefore needed to be put to the test;
- they were the four areas where unbiased observation would be difficult. I was concerned that the short observation time available might create pressure to generate data, thereby making it difficult to remain a neutral observer. Therefore to try and avoid unwarranted conclusions based on only limited observations I felt that using these headings would help the collection of 'suitably' focused and comparable data for analysis – a kind of triangulation;
- I believed those four areas to be important in terms of the National Curriculum requirements and therefore planning of work in the future.

A questionnaire was given before the children started their technology task, to gain a clearer impression of whether they understood exactly what was required of them in the task which was set. A further questionnaire was given later in the term to investigate my own supposition that the children only viewed technology work in terms of 'woodworking'. A final questionnaire was used to investigate the children's own evaluation of the work they had undertaken.

As headteacher I have responsibilities for the whole school in terms of management planning, curriculum planning, managing personnel, liaising with parents, governors, LEA officers, other outside bodies, etc. I have of course also the overall responsibility and care of all the children in the school. Fulfilling the role of headteacher I have approximately four hours per week, within school time to perform those particular duties when I am relieved of my role as class teacher by the part time teacher. During the Standard Assessment Task period, i.e. Spring and early Summer terms, up to three of the four hours was earmarked for work connected with Year 2 Assessment, during which I substitute for the Infant class teacher whilst she performs the necessary assessments.

In a small school environment the staff, in particular the headteacher needs to be flexible and to be constantly prepared for all eventualities. Visits such as that of the school dentist or doctor can cause disturbance not just to one class but to both. In the case of this school, limited classroom and storage space necessitates reorganization when such visits occur.

Thus the technology storage room, which also serves as the cookery room, photocopier room, music room and groupwork room also has to be used for any outside visitors.

With only six hours' secretarial time per week, phone calls must be dealt with at all times of the working day including teaching times. Visitors to the school, e.g. delivery men, repair men, maintenance staff, LEA officers, etc. all visit at all times during the working day and must be dealt with accordingly. Interruptions to normal class teaching times can be numerous since all those visits occur on an irregular basis, and are never 'booked' into the school diary. There are other visits, such as that of an assessment moderator, which are known beforehand and thus in that sense are less disruptive. However they can equally cause much reorganization of the normal working timetable.

IMPRESSIONS

Analysis of my classroom observations demonstrated that paired, single sex working was favoured by the children. Friendships seemed to be the most important criteria for the children choosing partners. It would appear from the observations that my role as 'teacher as helper' and as source of knowledge within the classroom was replaced by the children's use of each other when assistance was required. I appeared to have become the solution provider only when all other means of solving problems had been exhausted.

Observation confirmed that the making stage was definitely viewed as the most important aspect of all technological work. Plans were ideas in the head. The children seemed either unwilling or unable to generate more than one basic idea as an answer to a task, although ideas sometimes changed and developed as work began. Watching and talking to the children seemed to indicate that although they said they understood the value of design work, in fact they didn't really believe in it. All the guidance and record sheets, designed to help the thinking stage and create a record of their thoughts, were universally unpopular. Any writing connected with technology work was seen as a waste of time. Discussion with the children as to if and how the sheets could be modified in order to make them more useful would need to be undertaken.

Observations confirmed that many of the children had developed 'untidy' habits. These pupils left others cleaning away their materials for them. A sense of responsibility for the safe keeping of the tools and resources in the classroom was certainly lacking. It highlighted the need to develop the good practice of tidying as you go along, rather than a mass clear up at the end of the session, and the need to teach the children to be responsible for their own materials and tools.

There seemed to be a need for the children to 'play' with their models. Perhaps my impressions that 'time was wasted' testing things over and over again does not take into account the need for children to feel familiar with different materials before they can make full use of them. This has caused me to reflect on my attitudes towards play in learning for juniors. Unfortunately this research does not offer the opportunity to explore this issue further but it does provide a matter for future work.

When responses regarding planning were solicited in interview, it appeared that most of the group saw the value of plans before 'making', but they all found the planning sheets (designed to help them) were unhelpful because the sheets created 'more work'. Further investigation revealed that all the group thought they could manage to 'make' models without producing plans/designs because they had the idea in their head. Therefore they 'knew' exactly what they were trying to do. This leads me to suggest that the children don't realize that they did not know exactly what they were making before they started. They did not recognize that this was probably one reason for their difficulties in trying to evaluate their finished work, i.e. it couldn't be evaluated against a clearly described set of aims.

The interviews provided data of children's appreciation of the need for judgement and management of time. The most common way of estimating the amount of time needed to complete a task was to base it on the length of time available. Only two boys tried to estimate by assessing the length of time required to complete various parts of their model, i.e. apportioning time within the project. Even at the end of tasks, no child could really explain how long their model had taken. The estimates which they seemed to use were mostly based on thinking back from the end of the session, and guesswork as to how long the session had been. The interviews, I believe, supported my earlier supposition that the children may have a lack of experience of time management and awareness. They were not used to predicting time in ways which helped them to plan a finishing time effectively. Nor were they involved in making such judgements as the overall allocation of time.

Perhaps the emphasis on getting the work finished was now hampering progress in the use of designing and in producing quality work, since both aspects are viewed as time consuming. This appeared to be one issue which requires detailed further consideration.

One of the major problems identified by the children associated with 'making' was the misuse of glue. Another common strand was for the need to teach the skill of sticking various materials together. They provided little evidence that they realized that they lacked the skills of manufacture. 'Cutting wood' featured as a skill which some of the children would like to improve, which demonstrated that they did realize themselves that the accuracy and finish of work could be better. It also points to the fact that what they see as key skills are perceived as lacking. This could influence not only the actual quality of work and the length

Effective completion of technological tasks 45

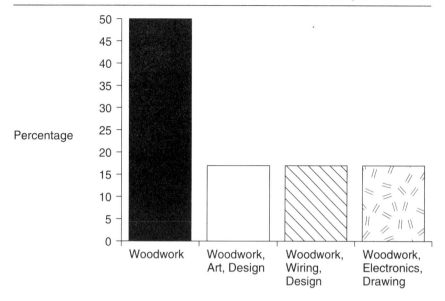

Figure 4.1 The aspect of technology the children enjoyed most

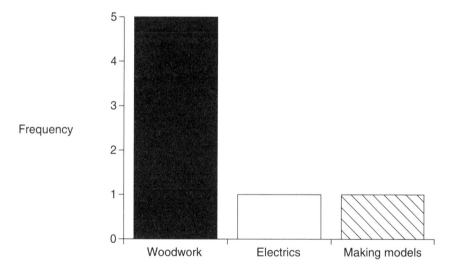

Figure 4.2 How the children defined technology

of time it takes the children to complete a task, but also their attitudes towards it. That could amount to a double bind for the children: recognition that their skills are not adequate to meet the 'standards', and that there isn't time to improve them before a task has to be finished. It appeared that a great deal would need to be done to further raise awareness of the need for improving technical skills, to invest in strategies for developing them, to engage children in evaluating skills use, and perhaps to involve them more in assessing their own practice.

The data seemed also to dispel the possibility that lack of communication between myself and the children was causing problems, since all but one child was sure they understood what they had to do. This seemed to be supported by the fact that no child wanted to discuss the project with me after it had been set. Six out of the seven children did not want any help before they started. One child wanted help designing.

In answer to the questionnaire item about what technology work was, two of the children stated simply 'woodwork'. Of the seven children, six mentioned woodwork in their list, and the one child who didn't, when asked to define his 'technical skills', cited the use of woodwork tools. Making models was referred to by three of the children. Design was mentioned by four. The aspect of technology most liked by this group was 'woodwork', with five children stating this fact. Two of the seven used a broader 'making things'.

The supposition that the children seem to view technology as 'woodworking' appeared to be justified. Likewise the supposition that as far as the children are concerned the whole emphasis of technology is on 'model making' was also reasonable. Analysis of the data conclusively points to an emphasis on 'making', particularly 'making things out of wood'.

'Lack of neatness' was the overwhelming reason for my not being satisfied with their work. This emphasized the information collected from the questionnaires which showed that this group of children definitely had a clear idea of one aspect of what I expected from them, with regard to the quality of finished artefacts. This was a piece of information I did not expect to collect. I felt the children didn't understand what was expected of them, or that if they did, they didn't really care. This was obviously a wrong supposition. However the revelation means that this issue needs to be investigated more. It confirmed the importance of questions as to how the children can be helped to satisfy my requirements and incorporate into their own thinking, and satisfy, concerns for manufacturing skills.

All the children thought the other children thought their work was good. 'Neatness' and 'worked well' were given as the reasons why the children thought their work good. The children were satisfied on the whole with their work and were happy that their peers agreed with their evaluation, in spite of the information I had obtained in the interviews. I feel

Effective completion of technological tasks 47

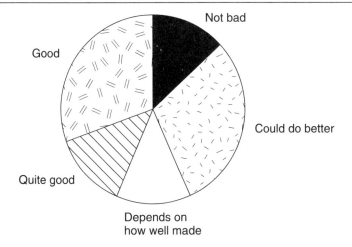

Figure 4.3 The kinds of measures the children expected me to use in relation to their technology work

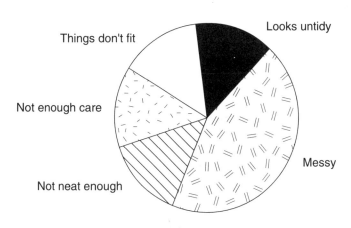

Figure 4.4 The evaluation criteria they thought I used

perhaps that although the interview information probably referred in particular to a specific piece of work, I was able in the interview to press the children to think more clearly about their work than in the questionnaire. Some aspects of my teaching came through on the questionnaire replies, e.g. my consent reference to planning work and taking care to produce 'neat' and 'tidy' work. Although the children seem certain that they knew what I looked for, they seemed unable to achieve those requirements. They also seemed less unhappy about this state of affairs than I was.

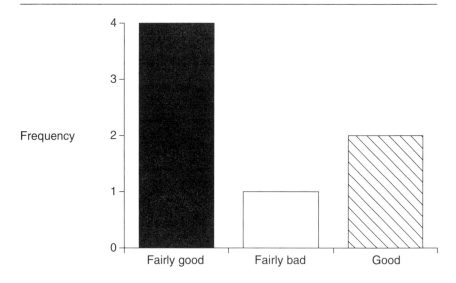

Figure 4.5 The children's evaluation of their own technology work

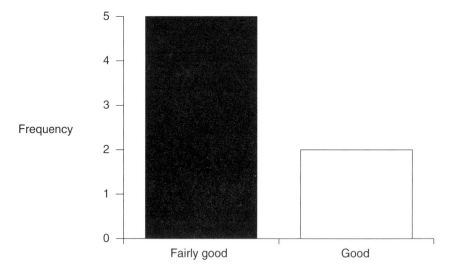

Figure 4.6 The evaluation they thought others would use

The evidence gave me a renewed perspective on the problem. My commitments remained to high quality work finished within appropriate allocations of time, and achieved by careful planning, followed by skilled execution of the plans. But there were differences between our perspectives. The questions which came to mind now were somewhat more instrumental and related to my teaching aims more directly than the earlier, speculative ones:

- does emphasis on time create problems with the attitudes to planning and quality of finish?
- how can the children be encouraged to value planning?
- how can the children be equipped to meet the 'quality' expectations?
- how can the children be helped to evaluate work more rigorously and critically through recognized criteria?

Although the initial question posed has not been fully answered, aspects of the problem have been more clearly defined as a result of the research undertaken so that further specific investigation can be continued. Some reasons I perceived as contributing to the problem have been eliminated whilst others still require more attention. The interview work has drawn to my attention to just how important it is to devote time to discussing the children's work in detail with them. It has thus made me focus on finding ways to achieve this in my teaching strategies, by engaging the children more fully in appreciating all aspects of the design process.

FURTHER READING

Department for Education (DfE) (1995) *Design and Technology in the National Curriculum*, London: HMSO.
Design Council (1987) *Design and Primary Education*, London: The Design Council.
Elliott, J. (1991) *Action Research for Educational Change*, Milton Keynes: Open University Press.
Tickle, L. (1990) *Design and Technology in Primary School Classrooms: Developing Teachers' Perspectives and Practices*, Lewes: Falmer Press.
Williams, P. and Jinks, D. (1985) *Design and Technology 5–12*, Lewes: Falmer Press.
Wisely, C. (1988) *Projects for Primary Technology*, London: Foulsham.

Chapter 5

Mixed ability children and the single open-ended task

Helen Deacon

Over recent years my colleagues and I have been introducing technology activities into our theme work, in order to ensure that all the children have experience in designing and making. Although we have mixed ability classes, which include statemented children with learning difficulties, we do not differentiate the technology tasks we set. Instead they are designed by the staff so that the children can approach them in an 'open-ended' way, at their own level of competence, using and developing the skills they possess and ultimately completing the work within their different levels of attainment. This approach is advocated by Newton and Newton (1991: 66) and Johnsey (1990: 8).

I was concerned that perhaps this approach of a single, open-ended task was not meeting all of my thirty 10 year old children's technological needs. Their models often seemed to be very simple and similar in terms of appearance and construction methods, use of techniques, and quality of finish. This apparent lack of technological breadth, or differentiation, in their work could be caused by a variety of reasons. I therefore began by listing some of these possible reasons under three key headings – children, teacher, task – and I hoped that my research would identify which was responsible for this lack of differentiation.

THE CHILDREN

- Were the more able children being too easily satisfied with the objects they were making and consequently not developing them as far as possible?
- Alternatively, were they unable to go further because they had insufficient technological knowledge and experience?
- How much group interaction was there which encouraged class answers to the technology problem rather than individual or small group solutions?

MY ROLE AS TEACHER

- Was I subconsciously leading the children to similar outcomes by my presentation and discussion of the problem, by the materials and equipment I provided the children with, and by my interactions with the children during the design and making stages? In short, were the children simply completing a teacher conceived solution?
- Was I giving the special needs children too much help, thereby raising their apparent attainment level but denying them the opportunity to investigate and succeed in their own right?
- Were my expectations of the more able unrealistic, based on my lack of understanding of their real technological knowledge?
- Alternatively, was I truly appreciating the complexity of skills and processes undertaken by the children during the design and making stages of the task? Was I tending to over emphasize the importance of the end product and was this a useful indicator of the processes and learning opportunities experienced by the children?

THE TASKS

Were they appropriate for the different ability levels in my class, or too limited and constraining and therefore stifling the imagination and creativity of the children?

The technology task I set in order to investigate these three areas was to 'design and make a scaled piece of furniture for the future, to fit a pipe cleaner person'. The children understood that after they had designed their furniture on paper they would be making a prototype out of art-straws before starting in wood. This, I hoped, would give them an opportunity to evaluate its effectiveness before beginning it in wood, and also give them a model to use for the dimensions they would need to cut out of square section and dowel.

In order for me to compare the differences in approach between children of perceived different 'abilities' I chose to study a 'more able' pair (Hannah and Laura) and a 'less able' pair (Gemma and Stacy) whilst they completed the furniture design task. Hannah and Laura are both very communicative girls who enjoy opportunities to work independently. They are extremely self motivated and high achievers in all of the curriculum areas. Gemma and Stacy are less high achievers, particularly Gemma who has language and mathematical difficulties.

I collected data from the five major stages of their work, i.e. design, making the straw prototype, evaluating the prototype, making it out of wood and evaluating the wood model. I decided to audio-tape and video each session and later analysed these tapes in terms of the methods and processes the children were illustrating whilst completing the task.

Data was also available at the end of the project from their completed evaluation sheets, their write-ups about the technology tasks, and through interviews with them. My technology diary helped me analyse my role during the design and making stages of the model. In this I recorded, as soon as possible after the lessons, my impressionistic and instinctive thoughts about each session and my judgements about how I felt my groups had worked. This helped me to highlight any mismatch between my impressions of and assumptions about what was happening and what was recorded by the means of audio- and videotape.

STAGE 1: DESIGN

In my opening session we speculated about furniture of the future, and the children were full of ideas, largely based around the concepts of gadgets and convenience. By now, the children were very keen to begin work, so after I had suggested they design two or three different ideas before agreeing on a final design they eagerly began work.

The two research pairs started immediately, but approached the task in very different ways. As can be seen in Table 5.1 the quality of discussion was far more complex between Hannah and Laura than between Gemma and Stacy. Hannah and Laura were full of ideas and before putting pencil to paper they freely discussed and explained these ideas to each other. There was no dominant partner and although Hannah put forward slightly more original ideas (23 compared to Laura's 19) Laura asked more directing questions to keep the discussion going.

They disagreed with one another (Hannah 8 times, Laura 14 times) but when an idea was rejected it was not pursued by the one who had suggested it.

Gemma and Stacy discussed fewer ideas and did not discuss them at any great length. Compared to Hannah and Laura's final three firm ideas of what they could make (a magic wardrobe, a vacuum cleaner you could sit on to do your cleaning, and a chair which feeds you) Gemma and Stacy only produced one idea of a spinning chair. Indeed, Stacy had begun drawing this idea four minutes into the session with Gemma closely copying her drawing.

As Hannah and Laura began to draw their ideas of a flying vacuum on paper, their discussion continued. They produced a number of rough sketches in their draft book before a mutually acceptable design was agreed upon and then they quickly drew a neat front and side view on paper.

Stacy, on the other hand remained in total control of her and Gemma's design. As she began to work out the circuit she would need to help the chair spin, Gemma sat playing with her pipe cleaner figure or pulling faces at the video camera. Once Stacy had finished her design she helped Gemma copy it exactly.

Table 5.1 Analysis of discussion, design stage

	Hannah	Laura	Gemma	Stacy
No. of questions asked by child to partner about task	2	9	1	3
No. of original ideas about design	23	19	6	13
No. of contradictions made by child against partner's ideas	8	14	1	0
No. of agreements	17	13	4	2

STAGE 2: MAKING PROTOTYPES

Primarily this stage was intended to help them gain experience in joining the art-straws with cardboard corners (a technique I showed them) before embarking on the square section wood model. I also wanted them to use their prototypes to work out the dimensions of their final models. From the video and tape recordings I made, I wanted to list and compare the skills illustrated by the two pairs, how they worked together, and whether they evaluated, discussed and changed their proposals as they met problems.

Since all the children in the class have had plenty of experience in using glue, art-straws and scissors, neither of my research groups experienced any difficulties constructing their models using these materials. They also confidently adapted to fixing the straws with the card corners. However, it was noticeable that whilst Hannah and Laura measured their art-straws carefully according to the lengths they had predicted in their draft book, Gemma and Stacy adopted a far more haphazard approach. Ignoring the lengths they had predicted they would need on their design sheet, Gemma began to cut the straws to any length. Stacy quickly stepped in and cut her a length to use as a measuring guide for cutting the other lengths she would need. Consequently, by the end of the session Gemma and Stacy had wasted many more straws than Hanna and Laura.

The most glaring contrast between the two pairs was the amount of discussion that went on during the session. Hannah and Laura spent a lot of their time debating different ideas. Indeed, when they began making their straw frame in this second session they had dismissed their original design of a flying vacuum cleaner and instead produced another design of a moving, feeding chair (their new design had been developed at home, so obviously their animated discussion had continued long after the tape had stopped!). Since this second design was not as clear or as detailed as their first they had to spend quite a lot of time explaining to each other

54 Helen Deacon

Table 5.2 Analysis of discussion, prototype stage 1

Skills and processes	Hannah	Laura	Total	Gemma	Stacy	Total
No. of disagreements with partner	14	13	27	5	3	8
No. of agreements	8	8	16	6	4	10
Asking partner question about task	5	7	12	2	1	3
Giving partner instructions/directions	25	19	44	5	10	15
Evaluating what they are doing	2	1	3	0	2	2
Observing a problem	6	6	12	0	0	0
Seeking reassurance from partner	1	1	2	5	3	8
Question about equipment/techniques	6	9	15	3	1	4
Explaining idea to partner	27	22	49	2	6	8
Showing object to partner	1	0	1	2	0	2
Interaction with others about task	5	7	12	1	1	2
Talking to partner about non-task	3	2	5	9	3	12
Interaction with others about non-task	7	7	14	4	1	5
Seeking teacher help	0	0	0	1	0	1

what they thought should be done. They were equally strong minded about the direction their model should take and consequently the conversation at times became a little heated. In total there were 27 disagreeing statements, 44 directions and instructions to each other, and 49 explanations about what they were doing (see Table 5.2).

Hannah and Laura were also aware of potential problems and difficulties, not only in the construction of the straw prototype, but projected the model into wood and evaluated it in terms of whether their ideas would be feasible in the different material. The headrest problem was a case in point. It began its life as a diamond shape above the back of the seat. During the first session Hannah and Laura had been concerned that the chair was too tall for their pipe cleaner man.

HANNAH It's [the chair] too big for our man. His head is supposed to be on the head rest.
HD What are you going to do about it?
HANNAH Leave it.
HD Fine.

Table 5.3 Analysis of discussion, prototype stage 2

Skills and processes	Hannah	Laura	Total	Gemma	Stacy	Total
No. of disagreements with partner	1	2	3	2	6	8
No. of agreements	1	6	7	4	2	6
Asking partner question about task	2	0	2	5	1	6
Giving partner directions/instructions	9	12	21	5	17	22
Evaluating what they are doing	0	2	2	2	3	5
Observing a problem	7	6	13	2	1	3
Seeking reassurance from partner	0	0	0	11	2	13
Question about equipment/techniques	0	0	0	6	0	6
Explaining idea to partner	4	3	7	2	3	5
Showing object to partner	0	0	0	2	0	2
Interaction with others about task	1	3	4	0	0	0
Interaction with others about non-task	0	3	3	4	2	6
Interaction with partner not about task	2	1	3	20	14	34
Seeking teacher help	0	0	0	2	1	3

After a while Hannah came up again.

HANNAH   No we're going to make the seat back shorter so it fits.

This seemingly simple decision on their part (as I had judged it during the session) had in fact been quite difficult to reach agreement upon. The tape revealed that although they both agreed that the headrest had to be lowered they could not agree on how to do it. Laura wanted Hannah to cut the rectangular back smaller. Hannah simply wanted to cut the headrest off and use it as decoration. In the end, after a friend's involvement and Laura becoming sidetracked by the length of the seat problem, Hannah became more dominant and took over completely. Rather than orally explaining, she demonstrated practically on the actual model. Laura appeared to approve:

LAURA   That's a better thing than I said.

However, the headrest decoration did not remain in the centre of the seat. During the second straw session it was disposed of completely because

Hannah and Laura felt that it would be too difficult to construct out of wood. Laura was also not sure about the aesthetic quality of it:

LAURA   I don't like that bit in the middle.

Stacy and Gemma contrasted quite sharply with Hannah and Laura's discussion (see Table 5.3). Whereas Hannah and Laura shared the decision making and contributed an equal amount of ideas and opinions, Stacy was again obviously the dominant partner. In the first session Stacy allocated them each a part to make, her the seat and Gemma the back, and they proceeded to make these parts with very little interaction. When Gemma had finished she simply sat waiting for Stacy to finish the seat of the chair, and to give her further instructions. She sought reassurances from Stacy 11 times, asked 5 direct questions about the tasks and was given 17 directions from Stacy. On the rare ocassions Gemma did offer an idea it was usually disregarded by Stacy.

GEMMA   I've had an idea ... when you sit on it.
STACY   Yes.
GEMMA   When you sit on it a little bit will hang down and you can put your feet on it.

Stacy gave no answer or reaction. Gemma did not pursue it any further. Both pairs decided to introduce a moving part. Gemma and Stacy to spin their seat, Hannah and Laura to turn the wheels of their moving chair. Both pairs had considered where they were going to put the battery and wires: Gemma and Stacey inside the base, Hannah and Laura on the back.

STAGE 3: EVALUATION SHEET

Before the end of the second art-straw session Hannah and Laura felt ready to complete the evaluation sheet about their model. They spent ten minutes completing it but in that time continued their discussion about what they had done with the headrest and began a new disagreement over whether they should also dispense with the chair arms. Hannah felt that they would be too difficult to fix onto a wooden frame and get in the way of the moving parts. Laura wanted to keep them declaring that it would improve the overall appearance of the chair.

LAURA   It'll look too bare without them.

They swiftly reached a compromise by deciding to leave the arms on to see what it looked like on their wooden model, and rushed up to show me their identically completed evaluation sheets.

   Gemma and Stacy's completed sheets were also identical to each other and I presumed that Gemma had simply copied Stacy's sheet. To an extent the tapes revealed this to be the case.

GEMMA   Stacy we're supposed to be writing this together.
STACY   You can copy.

Stacy had then given Gemma careful instructions about how to fill in her details, name, date, etc. However the tapes also showed that Gemma was more aware and personally involved in the development of her model than she had previously appeared to be. Rather than being totally dominated by Stacy she was beginning to offer some opinions.

GEMMA   (reading sheet)   Why did you change your design?
STACY   We didn't follow our design closely.
GEMMA   We did.
STACY   But we haven't put the triangles on there ... or the arms.
GEMMA   Well we would if she had given us more time ... if Miss Deacon had given us more time we could have put them on.

Gemma apparently felt quite strongly about this as it was a point she repeated in an interview later. She was also beginning to show an awareness of what the next stages of the model might be.

STACY   (reading sheet)   Do you think your model will work in wood?
GEMMA   Yes.
STACY   (reading sheet)   What materials will you need to finish the model?
GEMMA   Wood.
STACY   Material.
GEMMA   Wood, material ... sponge for the seat Stacy?
STACY   Yes.

Gemma was still seeking Stacy's approval but had for the first time, in the task, appeared to be more directly involved in it. However this involvement was to be short lived. When they were next required to measure their prototype for the lengths they would need in square section wood and dowel Gemma reverted back to dependence on Stacy with a dismissive,

GEMMA   I hate measuring ... I can't even measure this.
STACY   Don't do it like that, do it like this.

Stacy took control again whilst Gemma sat watching.

Hannah and Laura shared the responsibility of listing the necessary measurements. They checked each other's results and argued over millimetres. However they, like Gemma and Stacy, failed to use this evaluation time to consider where and how they were going to fix their mechanisms onto the model.

STAGE 4: MAKING THE WOODEN MODEL

Like most of the children in the class Gemma and Stacy approached the wood making stage of their design task with far less confidence than they had shown when making their chair out of art-straws. They began by seeking teacher approval for every corner they stuck with the hot glue gun, and Gemma was quite nervous using the hacksaw and bench hook, requiring a number of reminders about how to keep her bench hook secure on the worktop. Hannah and Laura, on the other hand, were far more confident and swiftly cut and stuck the pieces of wood they needed to complete their wooden frame.

In all other respects the wood making sessions closely mirrored the straw prototype lessons in terms of the skills and processes illustrated by the two pairs. Again Hannah and Laura shared responsibility for their model's development, giving almost an equal number of instructions to each other (Hannah 22, Laura 19, see Table 5.4). Each continued to show her partner what she was doing and swiftly foresaw and articulated potential problems (Hannah 13, Laura 13). Gemma and Stacy also followed the path established during the prototype stage with Stacy firmly in control and instructing Gemma (Stacy gave Gemma 30 instructions, Gemma only giving Stacy 8). When Gemma was not given clear instructions she would either wait to be told or asked a direct question. Consequently they worked at a far slower pace than Hannah and Laura and by the end of the session had only got the seat assembled and the lengths of wood for the stand cut.

Gemma's lack of involvement was further illustrated in the next session when Stacy told her to assemble these pre-cut pieces into the stand.

GEMMA   How are we going to make the stand?
STACY   I don't know.
GEMMA   You made the stand and you don't even know how to make it ... right you do the stand Stacy 'cause I don't know how to do it. If you do it wrong, you do it wrong ... I'm not having nothing to do with the stand 'cause I wasn't ... eh Stacy.
STACY   What?
GEMMA   We did do a stand didn't we?
STACY   Yeah I think we did.
GEMMA   Well I made the seat thing ... we did have a stand, but not like this.
STACY   We need another four bits.
GEMMA   Stacy I've got the four bits.
STACY   We need another four bits.
GEMMA   You're going crazy ... I don't know how I'm going to make this stand ... I do not know. This is a stupid idea isn't it Stace ... to make this stand.

Neither Gemma nor Stacy thought to go back and study their art-straw prototype to refresh their memory. Its usefulness was disregarded by them both.

Mixed ability children and the single open-ended task 59

Table 5.4 Analysis of discussion, making the wooden model

Skills and processes	Hannah	Laura	Total	Gemma	Stacy	Total
Disagreements with partner	6	6	12	7	2	9
Agreements with partner	4	5	9	1	1	2
Asking partner question about task	5	8	13	12	4	16
Giving partner directions/instructions	22	19	41	8	30	38
Evaluating what they are doing	0	0	0	0	0	0
Observing a problem	13	13	26	0	0	0
Seeking reassurance from partner	0	0	0	6	0	6
Question about equipment/techniques	12	14	26	1	5	6
Explaining idea to partner	25	17	42	5	6	11
Showing object to partner	10	7	17	1	0	1
Interaction with others about task	8	6	14	2	1	3
Interaction with others about non-task	3	4	7	4	1	5
Talking with partner about non-task	2	3	5	8	3	11
Teacher reassurance	18	20	38	7	6	13

STAGE 5: ADDING MECHANISMS

Having swiftly completed their wooden frame, Hannah and Laura now had to face the problem of making their chair move. All of the children had had some previous experience of building an electrical circuit during our machines topic, when we had built 'Marvellous Machines' using battery powered circuits which either lit up a bulb or sounded a buzzer. So Hannah and Laura began quite confidently collecting together battery, battery holder, motor, wires and wheels, thinking that these, connected together, would power their chair.

After a while they came up to talk to me because they could not agree on how to fix the pieces onto their chair. I discussed with them a variety of methods to power it, including battery and elastic band power. Hannah, using her experience in Year 4 of making elastic band powered toys, immediately went off to investigate the elastic bands whilst I showed Laura, using a drawing, how to set up a drive pulley linking the axle and motor. She went back to her place and immediately dismissed what

I had shown her and instead began to fix the wheels directly onto the motor.

Later Hannah came up saying that she could not get the elastic band power to work. We looked at the cotton reel she was using. She had threaded an elastic band through the centre of the reel and had fastened a small piece of match to one end of the band and a short piece of dowel to the other. She had then twisted the elastic band up, using the dowel as a handle, but when the dowel was released it did not spin as freely as she had expected. She was able to identify the problem: friction between the rough sides of the reel and the dowel, and even offered a possible solution, adding wax to the side of the reel. However she was not interested in pursuing this line of development any further, preferring to help Laura with the battery.

During the rest of this session, and into the next, Hannah and Laura worked continuously on their mechanism. Their major problem was that their wooden frame was too small to support battery holder, battery, wire and motor. They soon recognized this and attempted to give themselves more room by taking the arms off their model. They also realized that simply putting the wheels onto the motors was ineffective as the wheels kept spinning off. By the end of the session they had therefore fixed a small cardboard box onto the base of their seat, to give them greater space for the equipment, and added axles to their wheels.

At the beginning of the next session they sought my help to get the battery to spin the axle. I provided them with a drive pulley screwed onto the motor and having explained, through drawings, the principle of the drive pulley we again attempted to get the system to work on their chair. However, at the end of this second mechanism session and with various avenues explored we still had not succeeded, and consequently all three of us were feeling very frustrated.

The mechanism part of Gemma and Stacy's seat also caused problems. Stacy was by now clearly becoming exasperated with Gemma and her remarks were increasingly sharp. As Table 5.5 shows, she gave Gemma far fewer instructions than before (8 compared with 16 when making the stand). She also explained her actions less and began to seek my help more. Gemma withdrew even further, watching rather than participating. Stacy collected the electrical pieces she needed and covered the stand in card. She was able to get the motor to work on its own, but could not fix it onto their seat, and came up to me for help.

HD Who remembered how to fix the circuits?

GEMMA [pointing to Stacy]: I didn't know ... she just did it.

HD OK, well you'll still need to put something on the base of the seat to give your motor something to spin on. [I leave]

GEMMA You'll do that ... Stacy won't you?

Table 5.5 Analysis of discussion, adding mechanisms

Skills and processes	Stand		Mechanism		Total
	Gemma	Stacy	Gemma	Stacy	
No. of disagreements	4	2	0	0	6
No. of agreements	0	0	0	0	0
Asking partner question about task	8	1	7	0	16
Giving instructions/ directions	2	16	3	8	29
Evaluating what they are doing	0	0	0	6	6
Observing a problem	1	2	1	5	9
Seeking reassurance from partner	6	0	5	0	11
Teacher help/reassurance	3	3	2	7	15
Question about equipment/techniques	0	0	0	0	0
Explaining idea to partner	2	5	0	0	7
Showing object to partner	2	2	4	0	8
Interaction with others about task	3	6	1	5	15
Talking to partner about non-task	6	2	7	1	16
Interaction with others about non-task	2	1	6	2	11

After Stacy had covered the seat she came for help to get it to spin using the motor. We were unsuccessful. Several pupils made suggestions: adding Plasticine to connect the motor and seat; using additional batteries; fixing dowel from the seat to the base, and using string to pull it round; adding a cotton reel to give the seat a greater area to stick onto; anchoring the dowel into a yoghurt pot of sand. While Gemma remained apparently confused, Stacy told her to cover the seat and paint the base, while she organized the dowel, cotton reel and sand. With this peer help they finally completed their piece of furniture for the future.

STAGE 6: AFTER THE MOVING CHAIR

After the frustration of their chair episode, Hannah and Laura arrived at the next technology session equipped with a completely new idea. Annoyed that they had been unable to show the rest of the class a near completed model during the 'showing' session, they were now fuelled with

Table 5.6 Analysis of discussion, multi-storey model

Skills and processes	Hannah	Laura	Total
No. of disagreements	3	5	8
No. of agreements	10	10	20
Asking partner a question about task	20	9	29
Giving instructions/directions	15	12	27
Evaluating what they are doing	0	0	0
Observing a problem	1	1	2
Seeking reassurance from partner	0	0	0
Teacher help/reassurance	1	1	2
Question about equipment/techniques	13	9	22
Explaining idea to partner	14	23	37
Showing object to partner	1	0	1
Interaction with others about task	4	3	7
Interaction with others about non-task	3	5	8
Talking to partner about non-task	3	6	9

Table 5.7 A comparison of types of statements made by Hannah and Laura

Types of statements	Previous	This session
Disagreeing	12	19
Agreeing	9	20
Questioning	13	29
Explaining	47	37

ambition to produce something to fulfil the design brief. Neither were interested in pursuing their chair idea. Instead they set to work quickly, planning on paper Laura's idea of a multi-storey bed(room):

'It is to save space in houses of the future', she explained. Laura was more dominant in the initial stages of the model (see Table 5.6). There were fewer disagreements, a point recognized by Laura.

LAURA   We're cooperating today.
HANNAH   What?
LAURA   We're cooperating today aren't we?
HANNAH   We're doing it a good way.

A comparison of disagreeing, agreeing and questioning statements between their earlier wood making session (Table 5.4) and this session further highlights this degree of cooperation.

By the end of this session they had designed their model on paper, cut the wooden pieces and glued them to form the cuboid frame, had covered the sides with card and begun sticking the floors on. In their last session they completed the 'floors', painted the outside and added a lot of details in each bedroom (curtains, beds, sinks, washing machines, etc).

HANNAH So that you can do all of your household jobs while you are still in bed.

They had also agreed upon adding a light to the top bedroom and this time the circuit caused them no difficulty. Hannah quickly set up the circuit, including a folded tin foil switch, using the knowledge she had gained from our machines topic. Their revised model was now finished after just two sessions and they had finally completed the design brief of making a piece of 'furniture' for the future.

My research revealed that in my original practice, concentrating on the end product, I was overlooking the range of skills and processes exhibited by the children during the designing and making stages of their artefacts. This project showed me that both the processes and the products could be varied, in the ideas generated as well as in the quality of finish of work produced. However, providing support for differentiated responses and individualized needs was itself a problem. Solutions to the dilemma seemed no nearer, and reading ideas proposed by others confirmed that I was not alone. Lees (1984) suggests a complex system throughout the duration of the open-ended task whereby the teacher equips any child who needs help with resource information and instruction sheets at appropriate points of their work. This would obviously necessitate very tight organization and pre-planning and may, ultimately I think, allow the child to become too dependent on teacher answer sheets. Lund (1984) on the other hand recognized that sustaining the less able child's interest on a single project could be difficult, as was the case for Gemma. He therefore advocates an award system which encourages and rewards the child as they practice and learn new skills.

There were other matters which emerged which would also need thinking about. For example, it was quite clear from my data, especially from my personal technology diary, that I had not been as responsive to Gemma's needs as I might have been. Often I had recorded my satisfaction of how well Gemma and Stacy had worked, but when listening to the audio-tapes it became obvious I had misjudged their working together. I had rarely directly involved myself in 'their' task, other than the odd cursory remark and so had not intervened to prevent Gemma's growing alienation. A glaring example was with the straw frame prototype. I had assumed that they had completed it and hurried them onto the wood. Both the tape and Gemma's interview showed that they had not finished it. Perhaps this was why they appeared to be less confident when beginning their wood model. When I asked Gemma why she had not told me that they had not finished she damningly replied: 'I didn't think you would listen.'

Later, when she and Stacy sought my help with their spinning seat, my answers were too swift and short to be of any real help to Gemma.

Consequently as soon as I left, Gemma continued to sit looking at her chair, apparently unable to continue with making her mechanism work.

In order to involve all children 'technologically' in open-ended investigations, I will need to be far more responsive to their needs during the session, and I will only be able to achieve this through more vigilant monitoring of the factors I have identified. My research has shown that I need to:

- be more careful in my assumptions of each child's 'progress';
- increase my personal technological skills if I am to be an effective guide and helper to my mixed ability class;
- define tasks in ways which encourage the generation of ideas and support children's decision making;
- introduce skills, techniques and expertise with tools and materials in ways which help to extend the range of possible solutions to problems.

Only then will I be able to ensure maximum participation in designing and making by all pupils.

FURTHER READING

Department for Education (DfE) (1995) *Design and Technology in the National Curriculum*, London: HMSO.

Elliott, J. (1991) *Action Research for Educational Change*, Milton Keynes: Open University Press.

Hopkins, D. (1985) *A Teacher's Guide to Classroom Research*, Milton Keynes: Open University Press.

Johnsey, R. (1990) *Design and Technology Through Problem Solving*, London: Simon & Schuster.

Lees, J. (1984) 'Mixed ability approach to technology projects', *Studies in Design Education Craft and Technology* 16(2): 99–103.

Lund, D. (1984) 'Craft, design and technology for children having special educational needs', *Studies in Design Education Craft and Technology* 17(1): 10–16.

Newton, D.P. and Newton, L.D. (1991) *Practical Guides: Technology Teaching Within the National Curriculum*, London: Scholastic.

Tickle, L. (ed.) (1990) *Design and Technology in Primary School Classrooms: Developing Teachers' Perspectives and Practices*, Lewes: Falmer Press.

Williams, P. and Jinks, D. (1985) *Design and Technology 5–12*, Lewes: Falmer Press.

Wisely, C. (1988) *Projects for Primary Technology*, London: Foulsham.

Chapter 6

Views and values

Sarah Humphreys

The impetus for this research came from my asking two questions:

- What criteria do the children in my Year 6 class use to evaluate manufactured electrical toys?
- Do they use the same criteria when evaluating the electrical toys made by themselves and other members of the class?

These questions came from speculation that while they might be used to judging manufactured toys, the children were having difficulties in establishing and learning a whole new set of criteria in order to evaluate their own products.

Making judgements about manufactured products is an explicit aim in teaching design and technology. To encourage children to closely observe and develop qualitative opinions on existing models and designs, the national curriculum requires that pupils should be taught:

> to investigate, disassemble and evaluate simple products and applications ... relate the way things work to their intended purpose ... how materials and components have been used to meet people's needs and what users say about them ... to distinguish between how well a product has been made and how well it has been designed ... and to consider the effectiveness of a product, the extent to which it meets a need, is fit for purpose and uses resources appropriately.
> (DfE 1995: 5 paras 5f, g, h and i)

In this study I asked my pupils to evaluate familiar products from the viewpoint of a user – to verbalize, or in some way expose, their views and values as users of manufactured electrical toys. I had to presume that in order to have a 'view' – to express an opinion on, or a preference for, a toy – the children would be using some criteria on which their views would hang. To be able to express a view suggests a reasonable conclusion emanating from a concrete thought or belief. It is these 'concrete' assumptions (e.g. It's good because it goes fast – fast equals good) that I was defining as being criteria. In the example I've just given, 'speed' would be the criterion. I felt that the criteria used to reach these conclusions ought to be relatively easy to identify. What is a 'value' though? My understanding is that a value is something more deep-seated, more intuitive – less easy perhaps to define. I was aware that I might not recognize evaluations based on value judgements because there were no clear criteria to pick up on. However, the reply 'I like it because I do' could not be ignored or irritably dismissed on my part on the grounds that it was a comment which lacked a focus and could not be categorized. I presumed again that there must be criteria behind such value judgements. I did not however presume that it would be easy to define them.

I was particularly interested in whether, when making judgements about the manufactured toys, the children would be able to relate the ways things work to the processes of design and manufacture. I imagined that although they might be able to assess the desirability of the materials used to produce a manufactured toy, the leap from the processes used to manufacture the toy to whether the toy was 'good to play with' may be just too great for them to consider. This was interesting to me because I also imagined that when it came to evaluating their own toys this problem of holding views and articulating underlying values would be a major factor in their assessments of whether the toy was a success or otherwise. This brought me back to the new design and technology Programmes of Study, where it is recognized that evaluation is an integral part of the whole design process. Children are expected to make qualitative judgements about their own designs and products at various stages:

> evaluate their design ideas as these develop, bearing in mind the users and the purposes for which the product is intended, and indicate ways of improving their ideas.
>
> (DfE 1995: 4 para 3g)

However, for the moment my focus was on how the children evaluate finished products rather than how evaluation fits in with the designing and making of a product. The criteria involved could be similar for both sets of toys (manufactured and child-made). Although plans and illustrated designs were used in the process of making both, they weren't included in the final evaluations of others' work. The design was what they saw in front of them, not necessarily what the maker of the toy had intended. For the purposes of this study I interpreted 'quality of manufacture' to mean how well the toy was made or constructed.

For the purposes of organizing and later comparing the criteria the children used for the two types of toys I placed those which they initially displayed under one of three headings, namely, those which took into consideration the function, the appearance and the technical quality of manufacture of a toy.

Data were gathered in a variety of ways: questionnaires, videotape recordings, audio-tape recordings, and notes of the children's comments. I decided to have the children construct their own questionnaires, to give them a starting point for discussion, and as a way potentially of revealing the criteria the children might use in their evaluation of manufactured toys. I categorized the data/criteria which they included in their questionnaires into the three groups:

Function

    noises that are made
    buttons/switches
    if the toy moves
    whether it has wheels
    if it is fast or slow
    if it has moveable parts
    if it has flashing lights
    if it works by remote control or has a switch

Appearance

    the colour of the toy
    the size of the toy
    whether it has 'stickers' on it
    how 'expensive' the toy looks
    if it has a 'nice' shape

Manufacture

    the material the toy is made from
    whether it has to be built by the child
    if it can be taken apart and put back together again
    whether the toy looks safe
    whether it looks strong and as if it would last a long time

I asked the children to play with a variety of electrical toys, commenting on whether they liked them or not, pointing out what was good or bad about them. They took it in turns to demonstrate each toy, and they also used the questionnaires to help them. I collected information by video, listening for any criteria the children used. I have summarized my observation in the lists of criteria (below) that add to the lists that came from the questionnaires. One (in brackets) from the questionnaire was not mentioned on the videos.

Function

>   the direction in which it moves
>   whether the toy communicates verbally
>   if the noises are 'musical'
>   'bits' that can be taken off the toy
>   if it can be taken apart
>   whether the toy uses batteries
>   (if it has flashing lights)

Appearance

>   whether the wheels were big or small
>   if they like the name of the product
>   'is it worth the money?'
>   if the toy looks decent/boring/cool/doodie

Manufacture

>   if it has been broken and mended
>   whether it is easy to 'control'

In a further session I split the class into four groups of six and had them voice their opinions about a set of toys made in class. I sat in with each group and recorded the criteria, both on tape and on a chart. An example of the comments follows. Group three were looking at a tractor/truck with big wheels and four lights.

M It happens to be a very nice car.
K Yeh, it's sort of tractor type thing isn't it, really, HOT SHOT or something it says. J— made this. I think the planning that went into it is immense, but I think it should've, it was good in theory but not in practice.
O Um, well, OK, it's different, what is it, it's a sort of thing with big wheels which says HOT SHOT with bars and turbo arrows. Eh, it looks as if it's war because its red.
M It's like K— said it was good in theory but not in practice. If he had have folded the thing [M is talking about the card covering the vehicle] and put them on it, it would have looked better, and spray painted it like mine, but the thing is it's a bit short and ...
N It doesn't look much like a tractor, it looks like a truck with a thin cap, doesn't it. The back's too short for instance, they did try to have four lights on.
J I think it's good, but at the back it looks a bit like a house, because it's got four windows like a normal house. I don't like the idea of the

big wheels and the door doesn't open, so my feeling is that it isn't very good.

Some children became totally absorbed in their playing and practically disappeared, others didn't seem to want to play with any of the toys, and were much happier just talking about them. There was, generally, a marked difference between the way in which the girls and boys in the groups approached the toys. Many of the boys involved themselves in 'silent' playing, or fiddled around with the toys while the girls talked. It was tempting to get sucked into the gender issue and forget about 'criteria'. But I stayed on task to discover that this evidence almost replicated the criteria in the previous lists.

The children had five weeks of technology lessons in which to produce their electrically working toys. They were given pretty much a free rein as to what kind of toy to produce. However, the school's technology scheme for Year 6 states that all of the children will make a Jinks frame at some point. I made the stipulation that each member of the group should use this kind of frame within the construction of their toy, and that it had to have some sort of electrical component chosen from ordinary lightbulbs, flashing lightbulbs, motors, buzzers and wire. They had a range of construction materials to choose from. A variety of toys was made: a lighthouse, a boat, games which used buzzers, a dive-bombing aeroplane game, traffic lights, dolls' houses, a helicopter, a clown, a cat, road works and various vehicles.

I felt confident that the children had full 'ownership' of the toys – they really had made them themselves. The relevance of this to the research was that my aim was to look at the criteria they used to evaluate electrical toys made by themselves and other members of the class, not toys made partly by the teacher. There were some children who did not finish their toys within the given time. Some, because they had embarked upon a mammoth project, and even with frantic lunchtime sessions, simply couldn't finish. Others didn't finish because they lost interest in what they were doing.

I made notes after I had spoken to the children. I found that mostly there was a reluctance amongst the children to make any judgements about their own work (as opposed to each other's). They were quite happy to present it to me for my evaluation, and wait for me to say something. I'm keen on children's self-evaluation – it's certainly not something new for my class to be asked what they think about their own work. Usually, though, evaluation is done in writing, in the context of editing their written work. I thought these new opportunities for oral evaluations would tell me something of their views and values in the new context – but I was disappointed. I wondered why this was.

I collected more data by asking groups of children to evaluate a further, large selection of toys they had made and tape-recorded their responses as they looked at, played with, and talked about as many of the toys as they had time for.

They discussed each toy, usually constructively, taking into consideration a variety of criteria on which to base their evaluations. An example of this follows. Group one are looking at a small car which has a light on the roof and a propeller on the back:

S   What does it do?
JA   It lights up and the thing goes round.
S   Oh, it's good actually. Put it on the floor.
G   But it won't move.
JA   That's good, you can use this in the summer if you're too hot.
JU   As a fan.
G   That doesn't work.
JU   As a fan.
G   Well, that does work, but not much air come off it.
S   He's worked hard on the electrics as it's being such a small thing.
G   You should have a 30 volt battery in this.
H   And he's also made the outside quite presentable.
S   I think he worked more on the outside than the things inside. The only problem is the light and the fan don't turn at the same time. If you turn the light on then the fan has to go.
JU   He's done his electrics well.
H   You mean the motor's done nicely.
JU   Very nice how the card's been cut out.

In this short conversation, the group discuss the functions, manufacture and appearance of the car, as well as pinpointing problems with it and hypothesizing about how it could be used. Before listing the criteria there are various observations I made whilst listening to the tapes that I would like to report.

It was clear that they had no difficulties evaluating each other's work. They made what I judged to be fair and reasonable assessments using a variety of criteria, and were able to comment constructively about the positive and negative features of the toys, and on how they could be improved. When it came to evaluating their own work, however, the story was quite different. If the maker of the toy or model was in the group, they seemed more than happy to present their product and demonstrate what it could do, but they would not be drawn into commenting in any sort of judgemental way about it. For me, this is a very important observation. Initially I saw this reluctance to evaluate their own work as perhaps being attributable to the confusion children might have about the teacher/pupil relationship. If that were true, then it would be logical to suppose that in discussion with their peers they would not display this same reluctance – but they did, so it must be for some other reason.

Views and values 71

I began to wonder if the reason why the children found it so hard to evaluate their own work has to do with society's demands and expectations. In our society we respect success, and we will bashfully admit to it with prompting, but we hate the braggart – it is socially unacceptable to crow about one's achievements. Similarly, failure can be seen as something shameful, to be swept under the carpet rather than as a learning factor which needs to be confronted and dealt with. On the assumption that children soak up our conventions and implicit values we cannot expect them to suddenly flout these conventions by asking them to 'boast' about their successes and 'reveal' their failures. As it stands, this argument is undeveloped, but it is certainly something that I must consider further.

Transcription of the discussion group of which 'S' was a member shows that from the beginning she reacted in an emotive way, perhaps wishing that the group would move on to something else. The other children in the group tried to ask questions and make constructive comments about the model, but S seems to want to deny all knowledge of it, dismissing any notion of improvement. Similarly, the group wants to make the toy work, but S doesn't give them a chance, claiming that somebody had broken it.

H   What one shall we got to next?
S   Oh, I hate that, oh no, it's horrible.
JU This green one?
S   I'm not good at technology. I want to smash it up.
H   What's it supposed to be?
S   I don't know.
JU A road works?
S   Well I don't know what it's meant to be, it started off to be where the traffic lights go on, but I had to change it.
H   What's that thing inside?
S   I don't know.
H   It's like a little man.
S   It looks horrible.
K   What's it doing in there?
S   I don't know.
H   So you're just going to sling it in the bin then?
JU Does it work though?
S   I broke it.
K   Yeh.
S   I think it does.
H   Do the electrics work?
S   It doesn't work because somebody (interrupted).
JA Is that the only thing that's meant to work?
S   Yeh, well I didn't put anything else on it, I didn't get time.

H   Well why didn't you finish it at home, make it a bit more exciting?
K   Yeh.
S   Neh, I'll just throw it in the bin, it's boring.
JA It would be good if it had moving parts that dig up the thing.

A contrast to the above example was the discussion group three had about K's model of a clown's head. K behaved in a way common to many of the children involved in discussing their own models. She happily showed the group all that her model could do but made no evaluative comment.

K   It's a clown.
O   Oh, it's doodie, what does it do?
K   The bow tie spins.
O   Oh I see. Oh, it's really, really good.
K   It should work, oh, actually I think the wire's come off. But the bow tie spins, look.
N   By hand power ... what I like about this thing is that more than one thing happens, the eyes light up and even the bow tie spins round and round and round. I think it's everso nice, in fact, it's got loads of things that happen like the nose lights up and the bow tie spins. Totally amazing. J?
J   I like the hair colour, it's different and I like the nose because it lights up, do the eyes light up as well?

It tended to be the case that children within the same group would have similar reactions to a toy. Quite frequently, though, different groups had different opinions or focused in on different criteria when making their evaluations of the same object. A transcription made of groups discussing the same toy at separate times illustrated this. Group one viewed the toy in a positive way, mainly focusing on how it had been manufactured. They appreciated the design, but they were more impressed by the carefully made Jinks frame and the way in which the electrical component (a simple bulb and switch) had been fitted. Apart from acknowledging that it was a black cat, this group didn't comment at all upon it's appearance.

Once group two had established the toy's function, a member of the group immediately focused in on the toy's appearance, saying that it should have a bit more colour. From that point onwards the group seemed to regard this toy negatively, they were unimpressed by what it could do and didn't mention how carefully it had been made at all. They did make suggestions about how it could be improved: 'I think you could have a little bit more colour. Yeh, and if it was some kind of Jessie the Cat or something, If he had "My Name is ...".'

In group three one of the children made a point of saying how much they liked the colour of the cat, that it was black with just one white ear, they thought it was 'quirky'.

My initial attempt to distinguish between a view and a value – a value as something more deep-seated, more intuitive, less easy perhaps to define, continued to bother me. Listening to the tapes involved listening to children expressing their views about the toys in front of them. It was these views which reveal the criteria they are using to make their judgements. Different values are attached to these views, however, and that is where the differences in perspective seemed to come from. With the cat, for example, group one placed a value on good craftsmanship whereas group two were more concerned with appearance. Or was this a matter of who set the focus and held the attention of the others? – a case of holding particular criteria, selected by chance, in the frame?

From day one M designed and wanted to make a vehicle which would, predominantly, be big and silver. To do this he had to make carefully several Jinks frames which would fit together to form the body, to cover the frames, and then spray paint the whole thing. Wheels and the electrical component were incidental as far a M was concerned, and although a working light bulb was attached to the top of the vehicle, the wheels never materialized.

The group acknowledged that M had worked hard on the Jinks frames, but had little else that was positive to say about this toy. In the same way that it was important for M that the vehicle was silver, H, above all, felt strongly that the 'outside' could have been 'nicer', 'more colourful'. Other members of the group had other priorities, Ja and Ju objected to the lack of proportion of the model, and argued that for something so huge one lightbulb was insufficient.

On the whole, the children had clear views about all sorts of aspects – function, appearance and manufacture. It was when they were evaluating the toys made by each other, however, that the values that they attached to particular aspects of these views became clear, as their personal values held sway over their capacity to consider the range of criteria which could be applied in more detached ways.

It was at this point that I sought to 'answer' the questions which provided the impetus for my research: What criteria do the children in my Year 6 class use to evaluate manufactured electrical toys? Do they use the same criteria when evaluating the electrical toys made by themselves and other members of the class? I had listed under the headings function, appearance and manufacture the criteria that I was able to see and hear the children actually using in their evaluations of manufactured electrical toys. Now I was able to compare these criteria with the criteria they used to evaluate the electrical toys made by themselves and each other. To do this I used the same method of organizing the data (listing the criteria under the three headings), but I categorized these lists further by adding three subheadings, of criteria used:

- only in the evaluation of manufactured toys
- only in the evaluation of toys the children made
- in the evaluation of both sets of toys

Function

Manufactured toys

> the direction in which the toy moves
> if the toy is fast or slow
> whether the toy communicates verbally
> if the noises the toy makes are 'musical'
> if the toy can be taken apart

Children's toys

> if the toy 'does' anything or not

Both sets

> if the toy moves
> whether it has wheels
> noises that are made
> if it has moveable parts
> 'bits' that can be taken off the toy
> if the toy has flashing lights
> whether the toy is suitable for a boy or a girl
> what age group the toy is suitable for

Appearance

Manufactured toys

> whether the toy has stickers on

Children's toys

> whether the toy is in proportion
> how near the toy is to 'looking like the real thing'
> if the toy looks individual – 'quirky'
> how much detail there is on the toy
> whether the toy has a name or not

Both sets

> the colour of the toy
> the size of the toy
> if they like the name of the toy
> if the toy has a nice shape
> how much they would be prepared to pay for the toy
> whether the wheels were big or small
> if the toy looks decent/boring/cool/doodie, etc.

Manufacture

Manufactured toys

> whether it has to be built by the child
> if it works by remote control or has a switch

Children's toys

> how well the Jinks frames have been made
> whether the cladding has been cut out and stuck on neatly
> how the different parts of the toy have been joined together
> where abouts the electrics have been placed
> how carefully the electrical components have been installed
> whether a lot of planning has gone into the designing of the toy
> how much time seems to have spent making the toy

Both sets

> the materials the toy is made from
> whether the toy looks safe
> whether the toy looks strong and would last a long time
> whether the toy has been broken and is now mended
> if the toy works
> whether the toy is easy to control

Differences in the criteria involving function can probably all be attributed to the inevitable limitations the children had in making their own models. For example, I'm sure that if they had the facilities and faculties for making their own toys change direction, this would have become a criterion for the evaluation of both sets of toys.

As regards the criteria involving appearance, there were two interesting differences in the way the children evaluated their own products. The first involved criteria based on how 'realistic' the toy was, and whether the toy was in proportion. I wouldn't have said that the manufactured bright green and pink tractor which had orange stickers on was particularly realistic, but realism wasn't a criterion the children used when looking at the manufactured toys. It would be interesting to find out whether, when the children were talking about their own models not being realistic or in proportion, they actually meant compared to other toys rather than the 'real' thing (e.g. a tractor).

The most marked difference when looking at this data is the extra criteria the children used when evaluating their own models with regards to manufacture. They had all been involved in the process of designing and making a toy and in doing so they had become, to a lesser or greater degree, 'experts' in manufacture. They could use their 'expertise' to make reasonably sound evaluations on the quality of manufacture of each other's toys. They perhaps didn't feel suitably qualified (although this was probably not a conscious thing) to make judgements regarding the more sophisticated production of the 'manufactured' toys.

My impression that children found it more difficult to evaluate their own products than to evaluate manufactured ones should have been clearer – by 'their own' I had meant 'their own and each other's'. One clear difference that emerged was the ease and enthusiasm the children showed when evaluating each others' products, and the inability or reluctance they seemed to have when evaluating their own. I started to make the tentative suggestions (mentioned earlier) as to why this might be the case (in terms of social attitudes and self-esteem). To highlight and to look into this difference will be very important, whether these suggestions are sound or not. If I want children's self-evaluation to become a valuable tool for learning, looking more carefully at this dichotomy may well lead me to a better understanding of why self-evaluation was so difficult for these children, and how I might help them overcome these difficulties.

REFERENCE

Department for Education (DfE) (1995) *Design and Technology in the National Curriculum*, London: HMSO.

Chapter 7

Copying

Rosemary Jackson

There is quite a high degree of confidence about design and technology among members of staff in the school, and there have been several INSET training sessions on the subject. A senior member of staff has taken responsibility for equipping each classroom with a toolboard and materials needed for individual class projects. The school has a large craft room with workbenches, pillar drill and bandsaw, and this is regularly used by individual classes as the need arises and the timetable allows. It is also used for claywork as part of the lower school craft afternoon once a week. This craft cycle comprises textiles, food, technology and art and design. Other technology work in the school tends to be cross-curricular, and the lower school classes also do extra projects as is deemed appropriate by individual teachers. As a Year 5 teacher responsible for technology in the craft cycle, I have tried as far as possible to link the projects with Year 4 and Year 5 science work on forces and energy, levers and pulleys.

I decided to undertake research on copying because I have observed it to happen very regularly within technology lessons – children glance across the room and are inspired by someone else's idea; out of two children sitting together, one may find it hard to get started and copy the partner's work. Given that we need to assess children on their capacity to be inventive as well as imitative, it is sometimes hard to know from whom a good idea originated in the classroom. Copying happens frequently in creative lessons, where there is much movement and discussion amongst the pupils. I had noticed that it often evokes heated reactions from those being copied. I decided to find out more about the phenomenon, and the pupils' attitudes towards it. I decided to gather my data from observing the children at work and making dated entries in a diary, as well as by holding interviews with individual children and some of the members of staff, and writing down their responses for analysis. I also held a discussion with my class to discover their feelings about copying.

The word 'copying' seems to evoke an initial negative response from children and adults alike. Human beings are both competitive and cooperative. Whilst we enjoy recognition for our individual creative achievements, we also recognize that we realize more by sharing our ideas in teamwork, especially when problem-solving. When we learn we are both mimics and explorers and creators. At a national and international level, scientific advances which lead to national glory and economic advantages are closely guarded secrets. Technological products which will earn industrial companies large sums of money are kept under wraps until patents have been applied for and 'protection' granted. Clothes designers 'guard' their ideas prior to every major show. It is partly human nature to build empires and as much financial gain as possible. It is only 'fair', the argument goes, that the person who has spent hours devising a new computer software package should be able to apply for copyright, isn't it? On the other hand, many agencies are sharing appropriate advances in technology to help people in developing nations to be self-reliant and to improve their way of life, such as techniques for building wells and making farming and basic industry more efficient. Advances in health care are also a matter of communal sharing and human ethics, though not entirely so. But ethics in industrial design?

Some nations have gained an unflattering image as arch-copiers of earlier versions of motor and electrical technology, and now lead the field with many of their products. Their education systems gear people up to be hard-working specialists who will be devoted to their employer, making items designed by others. Throughout manufacturing the international commercial world embodies the ethics of self-interested, capitalist societies, where it is both acceptable to copy, in order to maximize profit, and unacceptable to copy, to protect market share.

There is another social dimension to this problem, which has implications for teaching. Technology is necessarily a derivative activity in that each step forward builds upon previous inventions, using them as a starting point for an improvement or modification. For the consumer of design and technology, we have reached the point where it is almost not worth buying an electronic appliance because it will so soon be superseded by something better. In terms of the design and technology curriculum, the drive for creating 'new products' has led to changes in assumptions about teaching and learning. For example, Dodd (1978) reviewing the history of design and technology teaching in this country, comments on the contrast between the rigid training in skills at the beginning of the twentieth century, and the later reaction to this, with the recognition of the need for 'creativity' and problem solving. Galton, Simon and Croll (1980) infer the importance of this 'shift' in relation to the primary school curriculum in general:

> [O]nce the students had either been told or had found out one rule, the discovery group tended to do better at finding new rules for fresh situations. This suggests that, when talking about discovery

methods, it is useful to distinguish between two different processes; first, learning the method by which rules can be discovered and second, using these methods to discover new rules for solving unfamiliar problems.

(Galton, Simon and Croll 1980)

Children are expected both to follow and to lead; to initiate and to imitate; to follow rules and conventions in reproducing solutions to given problems, and to identify problems for which they will propose and test their own solutions. They are also expected to compete with each other, as well as collaborate. Elements of these features of social life in classrooms are likely to be found in the pupils' responses to lessons, and design and technology might highlight the problem.

Having been part of the craft 'cycle' groups for the past few years, I had made some observations about copying. Having taught the skills, demonstrated the available materials and set the challenge of design tasks, the first group of children in the cycle starts on each project. They have to generate ideas and often work quite slowly so that one or two may not finish. Depending on the time available, some of the finished products may be quite rough. Subsequent groups approach the project with ideas of what they think the ultimate goal is, particularly if the previous group's work has been on display. They tend to select the ideas they want to use from what they have seen, and may have been thinking about what they want to do for a few weeks. The teacher has by now become more *au fait* with the time scale of the project and also with what definitely does and does not 'work', and is able to discuss some common pitfalls with the children so that they do not waste too much time. A few children will come up with novel proposals, but most either directly copy or modify existing ones for their own purposes. Many children complete the whole project more quickly and often produce artefacts with a better performance. This is illustrated in Table 7.1.

The design brief for each group was: 'Design and make a vehicle powered by a rubber band. See how far you can make it go and how good you can make it look.' It was the first time I had undertaken this project. The children made a simple chassis using a Jinks frame, building in the skills necessary for this, and then designed a body using card. Workshops

Table 7.1 Results of distance test

	Average distance travelled	Furthest distance travelled	Didn't go at all
Group One – Oct	39cm	1m 19cm	6
Group Two – Dec	79cm	2m	0
Group Three – Feb	1m 55cm	3m 60cm	0

on making skills were held when the majority of the group reached the stage where this was necessary. The final stage was to discuss a fair test for the vehicles, to test them for distance covered and appraise their aesthetic appearance. Comment in April by Year 6 teacher: 'These models just got better and better – there's a real difference between this lot and the first one, and I'm not just saying that!'

I was surprised at the vehemence with which the children denounced copying and copiers! The word 'angry' was used several times in connection with being copied and betrayal of trust by friends. They were very perceptive in some of their comments, for example:

'It is bad if it is your idea. You wouldn't trust your friend any more.'

'You would be very angry and change your design in private.'

'It's not fair to have to share work if you don't want to.'

'You *have* to copy from the board to get things right.'

'Copying is bad if you don't understand what you are copying.'

'If someone copied you and finished first and got you into trouble for copying, would the teacher believe you?'

'It's unfair if you copy in a test and get a good mark, and it is also unfair if the person who didn't copy is being punished for having the same result as the copier.'

'You sometimes copy when you are working in a group and have to keep up.'

'Copying is OK if you've agreed to let someone copy.'

The last two thoughts provoked further reflection from members of the class, some of whom may have thought I wanted to hear condemnation of copying (although I had said that I wanted to hear all their ideas). However, these comments also refer to copying which is 'legitimate', and the first one presumes teamwork and collaboration (and possibly also acknowledges peer tutoring, where one member of a group gives another help with a concept or skill, or even with 'ideas'). The final comment implies that the idea held by the person is not very important to him or her – maybe it is a second-hand idea or one that has not been 'costly' to the person who thought of it, in which case copying is OK. What is also clearly stated is that it is important to have been asked permission to use the idea, and not have had it 'stolen'. These are fine-line distinctions, suggesting that some children at least are very aware of the importance of context in judging their response. As Cohen and Manion say: 'The irony of cheating in school is that in familial and occupational settings, the same kind of co-operative endeavours are generally considered morally acceptable and even commendable' (Cohen and Manion 1981).

The subtleties of such distinctions went even further, suggesting the coexistence of multiple attitudes towards being copied: 'I would be angry and pleased if someone copied me; angry because it was my idea, but pleased because someone thought it was good enough to copy.' What is the way to solve this kind of dilemma? The person being copied needs to be credited with having had the original idea, certainly. Much also depends on the personality of the copied person, and it is hard to prescribe hard and fast reactions to a situation when the classroom environment is so fluid in any case, depending on so many subtle and emotional interactions between people from different home and cultural backgrounds. It also raised the question again of just what 'creativity' is, since it was becoming clear that what I really meant by 'copying' was borrowing a new idea from someone without his or her permission to use it, where you might not otherwise succeed by yourself. New ideas are at the heart of being 'creative'. Dennis Child (1973) has tried to find the definitive use of this word, without success. However, he has culled the following ideas:

> Other generalisations about the personal qualities of creative men and women ... depict them as single-minded, stubborn, non-conformist and persistent in tasks which engage their imaginations. ... Risk-taking and venturesomeness with ideas appeal to the creative mind.... As one might have guessed, in general the higher the level of abstraction attainable by an individual, the more creative are his or her concepts.... Creative output needs time and effort.
>
> (Child 1973)

It would seem that a high degree of self-confidence, an ability to use divergent or lateral thinking, frequently returning to the information provided and trying another approach, are prerequisites for creativity. Child suggests that the classroom can provide a good environment for all children to be more creative, through activities in which teachers 'recognise and value their pupils' ideas' (ibid).

Creativity could also be seen as a way of successfully communicating new ideas to other people: we seem to be focusing on ideas as being the important thing. After all, as the children themselves said, you have to copy from the board to get things right, and in general what is on a blackboard is not an original idea but a means to teaching an idea, providing an answer to a given problem, or conveying information. You also have to copy skills in technology, usually demonstrated by the teacher, to acquire mastery of them. The children pointed out that copying someone else's answers or ideas is not, by and large, ethical, particularly at school.

Indeed I have 'fielded' some very disgruntled people who feel that they were illicitly copied.

The children's over-riding reaction to being copied without permission is one of anger. I observed several examples of this during the lessons in which the task was to build a car powered by an elastic band, using wood strip and card for the chassis and body. The first group needed a fair amount of prompting to see how the elastic band could be wound round an axle to store energy for the car. Some of the boys were rather disappointed that their cars were not moving very far, and one decided to fix the elastic band so that he could catapult his car a great distance – it worked, of course, but contravened the rules of the design brief. At once, a group of boys in his vicinity copied the idea and were busy seeing whose could go the furthest. The initiator protested loudly that the others had copied his idea without even asking permission. He wanted to be the only one to be seen to have this new idea. His protestations were nipped in the bud when he realized that the solution was not permitted, but his feathers had definitely been ruffled by the copiers. His reaction demonstrated the strength of those views reported earlier:

'It is bad if it is your idea.'

'You wouldn't trust your friend any more.'

'You would be very angry and change your design in private.'

'It's not fair to have to share work if you don't want to.'

However, this incident also demonstrates one reason why copying regarded as undesirable (by me and the pupils) was a simple, unrefined notion. In another group working on the same project, the children had no difficulty in seeing how to use an elastic band to store energy. Their problem was rather how to make it more effective. One girl produced a car which consistently went a great distance, and looked immaculate. She was available to help her friend, who was having difficulty making her car go at all. We looked at the finished car to see what might be making it go so well, and discovered that she had attached one end of the elastic band to one axle, and the other to a drawing pin on the chassis. That reduced the braking effect of the band rewinding round the second axle. Following an initial reaction of 'It was *my* idea . . .' she agreed that she wouldn't mind someone else using it if she was given the credit for it. The subtleties, complexities, and sensitivities seemed to be increasingly evident.

Just how complex and how potentially confusing for pupils this could be was illustrated in an incident when a Year 4 craft group was doing paper engineering, recorded in my diary:

> One girl indignantly claimed that the girl sitting next to her had copied her idea. The other girl had copied rather badly in my view,

so I was able to admire the original piece of work and saw what a good job she had made of it. I said that I always made notes about each person's work and that I would record that she had thought of the idea first. She left me obviously not mollified or convinced that justice had been done.

Despite the awareness of the subtleties by some children, the majority displayed very strong feelings about rights of ownership of 'intellectual property' and were quick to say if they thought it unfair that someone else should have copied an original idea of theirs. This may have been more acute because of the expectation that creative work like technology should be a rather personal expression of one's self. It may even be much more important to an individual than, say, a purely academic exercise like a spelling test, where the task is to reproduce a conventional 'product' in a standardized form.

There is a corollary to copying which stems from this. Competitiveness is an element which is very strong among some of the children. Some are so intent on achieving their own solution to a challenge that they will refuse to take on anyone else's ideas and may fail to finish a project. I noted in my diary the case of one boy in the Year 4 class periscope project: 'It appeared that he could not get to grips with the principle of periscopes at all, and rather than accept help from anyone, he gave up the challenge.'

Perhaps this particular boy was the 'independent type' who prefers to work individually on an activity which is dependent upon personal qualities such as self-reliance. An instance such as this suggests that strong adherence to the view that copying is wrong might lead to such situations. When children are devoid of ideas, they could otherwise gain stimulation, make progress, and learn. Learning may be seen as something which can best occur in a social setting, where ideas are exchanged in a cooperative way. The possibility of one person testing out ideas with those of others, where there is interaction which results in an amalgamated resource, could prove difficult for the pupil with strong views on copying.

It can for others, however, be a great morale-booster if someone else thinks enough of what they have done to copy it, or enough of their idea to adopt it, incorporate it and perhaps extend it. I have already cited the tension of being both pleased and angry at being copied. One boy actually advertised a good idea he had in order for other people to use it, as I noted:

We were making pop-up scenes for firework night and I had demonstrated two possible mechanisms for the class to use, when this normally quiet and self-contained boy (but who has considerable artistic talent) proclaimed that he knew another way to do a pop-up. He quickly wove two different coloured strips of paper together to make a zig zag which could be attached by one or both ends to the folded card. Several others gathered round to see how it was done, and nearly every scene included one or more of these zig zags, used in different ways to good effect.

There may have been more to this than was apparent. It is possible that he took the idea from a source that others would have seen, too. If it was not an idea original to him, it would not have 'cost' him so much to share it. On the other hand, it might have been a way of gaining the respect of others in a sphere in which he felt confident. Perhaps ideas that are in 'common ownership' are OK to copy and to share with others, in the same way that we acquire and demonstrate knowledge and skills to gain mastery of them. And perhaps passing them around is a form of social currency – not unlike teachers sharing ideas about their work.

From this investigation of copying, it is now clear that there is a subtle difference between imitating commonly held knowledge, copying 'legitimately' (from the blackboard, for example, or with permission) and copying someone's own, creative idea without permission. The latter is the one that caused the strongest reaction from the children. The children had a wider grasp of issues related to copying, and there seems to be plenty of scope for not being clear where it is appropriate, especially if they sometimes work individually, and sometimes in groups. It is important therefore to discuss different aspects of copying with the children so that they are aware of appropriate behaviour in lessons – and this would apply equally to those who tend to copy and those who are often copied. Meriel Downey suggests that judgements: 'are open to various kinds of bias . . . they can have serious repercussions for pupils and can affect their progress in school, their motivation and their self-image' (Downey 1977).

It could be helpful for children learning to make good judgements to distinguish between appropriate and inappropriate copying according to whether they are working individually or in groups, and on common skills/techniques or 'ideas'. Building a good team atmosphere is often part of the personal and social education programme. If we require the children to solve a design problem as a team, they will need to know that all ideas within the group are subject to common ownership, and that 'copying' will be essential in order to build up to the best possible solution. Brainstorming ideas at the beginning would help each group member to feel included, and to share the focus of attention.

There is an interface between group work and individual work where respect for each other comes into play. In a group it is important to respect everyone's contribution. When working individually children also need to learn to respect other people's feelings and ownership of their work. At the start of a project with individual outcomes, it would be a good idea to discuss how people feel when they have been copied without permission, to agree procedures for working – perhaps a rule that no-one will deliberately copy anyone else's work unless they have that person's assent, and assure them that their own, unaided attempt will be much more valued in that circumstance. There is a saying that imitation is the sincerest form of flattery, but this is certainly conditional upon the person being copied being happy about being copied, in the views of these children.

FURTHER READING

Boydell, D. (1978) *The Primary Teacher in Action*, London: Open Books.
Calvert, B. (1975) *The Role of the Pupil*, London: Routledge & Kegan Paul.
Child, D. (1973) *Psychology and the Teacher*, London: Holt, Rinehart & Winston Ltd.
Cohen, L. and Manion, L. (1981) *Perspectives on Classrooms and Schools*, London: Holt Education.
Dodd, T. (1978) *Design and Technology in the School Curriculum*, London: Hodder & Stoughton.
Downey, M. (1977) *Interpersonal Judgements in Education*, London: Harper & Row.
Galton, M., Simon, B. and Croll, P. (1980) *Inside the Primary Classroom*, London: Routledge & Kegan Paul.

Chapter 8

Change from rigid teaching

Sue Lusted

There are many schools of thought on the methods employed to facilitate children's learning. In my classroom I had always used a 'chalk and talk' and 'demonstration' teaching style, which I considered to be traditional, and which was the method by which I had been taught. I had not even seen the possibility of teaching in ways which allowed pupils to apply their new knowledge, or to explore ideas, as suggested by Jeffrey, who says:

> Teaching facts is one thing, teaching pupils in such a way that they can apply the facts is another; but providing learning opportunities which encourage pupils to use information naturally when handling uncertainty, in a manner which results in capability is a challenge of a different kind.
>
> (Jeffrey 1991: 64)

In teaching design and technology in my new situation I planned and started to carry out work with the pupils using my traditional approach. Each pupil made identical items in my 'food' and 'textiles' lessons. The children made no contribution to lessons apart from mixing and weighing the ingredients, or cutting and fixing materials according to pattern. 'Outcomes' were regarded as the main objective of the lessons. If a child failed to make a satisfactory product I felt dissatisfied with the way I had taught the lesson, as if I had failed to transmit an instruction or a skill sufficiently to ensure that all children succeeded to a 'desirable' standard. I saw my role as being to set a task for the children, directing their practical work at every stage and ensuring that an item had been completed at the end of each lesson. I would give the children a recipe or template, show them step by step what to do, and hope that each child would leave my classroom carrying identical objects. I had adopted the didactic method of teaching because I believed it had distinct advantages in terms of classroom organization. I could provide the exact materials required, in appropriate quantities or cut to size in advance. Skills could be demonstrated to the whole class. The children's rate of progress could be monitored. No one could make lasting mistakes, as these could be rectified at each stage. If a child worked 'too quickly' extension tasks could be provided, for example: a piece of scrap material would be given to the child who had completed a sewing task, to practice sewing skills and try new stitches, and be discarded later. If a child worked more slowly or had problems with a skill they could be given 'extra help' in order to achieve a satisfactory outcome. I believed that by controlling the outcomes all children, even those who lacked confidence or who had 'limited ability' were likely to respond to the security of being told exactly what to do and knowing what they were going to make.

The Year 4 pupils had experienced one term of my 'direct' teaching when I began the research. They made a placemat. Each child was given a piece of material, some thread and a needle. The children were then shown how to do running stitch and then a knot, then they made a mat. The results were all very similar, the only variables being the amount of stitching the mat contained and the colour of the thread used. I began to realize, on reflection, that this style of teaching might be inadequate for design and technology teaching. It was at this point I asked myself several questions:

- Did my teaching method benefit my pupils?
- For whom was the outcome important, me or the pupil?
- Was the product crucial?
- Was I teaching in a worthwhile way and what did my pupils learn?

I realized that each child had a different technological and design capability, varying academic abilities, and different interests and ideas. If they all made identical artefacts what did each child learn apart from prescribed skills? How could I evaluate work that looks the same, to allow for children's differing abilities and capacity for problem-solving? If the children were allowed to think for themselves and pursue their own ideas, would the child with less ability or confidence achieve an outcome at all? These were just a few of the questions which emerged as my thoughts developed and this realization grew.

The aim of my investigation was to see if a change in my thinking about teaching methods produced a change in the working practices in my classroom. I attempted to move away from my traditional/didactic teaching style to a more open-ended, problem-solving approach. The change of style would, I hoped, produce a change in the way in which the children work and what they learn. This in turn would allow them to produce more individualized products. It might be expected that some children would produce similar artefacts to others, but no outcomes should be identical as they had been when using the traditional teaching approach.

I attempted to give the Year 4 pupils an open-ended task within their project on 'flight', where I would not know exactly what they would make.

However, I felt uncomfortable with this, and I knew I would find it difficult to assess when to guide and support the pupils without imposing upon or unnecessarily influencing their activities. I would find it difficult to decide how far to direct or suggest alternative lines of approach if I saw a child 'failing'. Controls on the activity were created by the way the task was set, and the materials provided, as these influenced what they could make. The background information they were given by me on the subject, and that which they received from their class and science teachers, also might have influenced their ideas.

The task chosen for the investigation group was to make a model bird with moveable wings to illustrate their flight capabilities and mechanisms. This was a task that I thought would provide design opportunities, and allow the pupils to develop artefacts which encouraged their own ideas, planning procedures, and the acquisition of a range of making skills. The model would also provide opportunities for evaluation. My concerns remained. The dilemmas became acute. I was certain that in the light of all I had 'discovered' about myself I should move away from directing the children to my 'right' solution, but I was reluctant to let go. I devised a means of classroom observation, which I regarded as essential for providing good quality information about the group and their reactions, and my responses to events. I also used a questionnaire at the end of the project to solicit the children's views about the changes.

The initial design brief was to produce a working model of a bird's wing. This task did not offer complete freedom as I thought it might be inappropriate to make a sudden and dramatic change. It gave the group some guidelines in which way to work and allowed them to make some choices for themselves.

The group was puzzled about the change in the type of activity, and asked me about the nature of the 'start' lesson and its relationship to sewing and cooking. Their first response to the task gave me the impression that some had been stunned into silence. I judged that they would need some guidance so that they should begin their task with some clarity about what was required. I suggested they start by producing an illustration of a bird in flight.

After the first lesson, which took place in the homecraft area, I thought that the pupils felt restricted by the nature of the room and its resources. We moved to their own classroom, where equipment and materials for the models were placed on a central table. The children appeared to me to be uncertain and reluctant to get started on their project. We looked at the various materials and pictures I had provided, and talked about the 'Skyhunter' programme on which this task had been based.

Participation and involvement were slow, until one pupil produced a large bag of bird feathers from his desk, apparently brought in for topic work. He suggested we pass the feathers around and look at them carefully to see how they were constructed. This prompted some of the others to suggest that we could draw bird feathers and produce feathers of our own that could be used in the making of the models – an idea which produced a flurry of activity and work commenced. The children divided themselves into working groups. Some chose to work in pairs, others worked alone. There appeared to be a little conversation going on. However, there was some swapping of ideas between groups and individuals.

Drawings of the initial designs were completed in their books. Many had formulated their ideas completely by the third lesson, and several had approached me indicating they had brought materials to use to execute their designs. I had prepared the centre table in their classroom and it was full of materials, tools and equipment they might require to make a prototype design of moveable bird wings. The materials provided for the task were: Pritt Stick; Marvin Medium; ordinary and double sided adhesive tape; paper fasteners; scissors; hole punch; stapler – long and short arm; Art-straws; string; cord; paper; card; tissue paper; coloured gummed paper; wooden sticks; dowelling rod. Two large sacks of junk which consisted of various containers such as margarine/icecream cartons, plastic bottles, meat trays, kitchen roll centres, egg boxes and yoghurt cartons.

I listened to their conversations and initially I heard comments between themselves like 'What shall I do?', 'Will this be OK?', 'Do you think this is right?'. I began to think I had not made the task clear and this had resulted in their apparent lack of confidence in how to proceed. After an initial working period children came to me holding their work and said: 'This is the best we can do.' They had made quite impressive beginnings, I thought, and so I was able to enthuse and give encouragement. I did this by stopping the class and showing their work. Other children made a few comments and comparisons to their own work, and carried on, working mainly in silence, though I had not directed them to do so.

Gradually the children began to talk more and I spent time writing down what was said to me and what they said to each other. The lesson had gone much the way I had hoped it would. Many pupils had made good progress with their prototypes. The dialogue had been restricted to begin with, but I assumed this was because the children had not been encouraged to voice their thoughts in previous lessons. When I stopped the pupils to show the artefact being made by Shaun and Christopher, I had encouraged discussion, giving the class the opportunity to offer their own opinions. Immediately afterwards Hannah G asked Amy 'Are we doing this in rough?'. Perhaps I might have seen this as lack of confidence, as Hannah G is a child who needs encouragement, but perhaps she was searching for information about the rules of engagement. Christopher A quickly informed her that this was for real so it appeared to me that he understood the task, and that the idea of a prototype had not been appreciated by either of them.

At first I believed that the restricting nature of my previous teaching style had limited these children in their attempt at a problem solving activity. I thought that the children's failure to approach me and discuss their designs was due to their lack of confidence in making 'personal' responses to tasks. However after listening to their conversation and watching them produce their work I wondered if they were too busy formulating their ideas to talk. If they required extra equipment or needed to be able to use a new skill they soon approached me. My role as instructor was evidently still clear to them, but now it was 'by request' rather than by command. Some pupils demanded much of my attention and were given considerably more time than others this way. Some did not approach me at all: pupils like Shaun and Christopher, who nevertheless produced good 'outcomes'. I assumed that these pupils knew what they were doing and needed no help. I felt the 'less able' pupils were finding this new situation difficult, as they were required to do something for themselves. Perhaps it was for this reason they were continually seeking attention from me and their peers, having developed an expectation that 'help' would be given to produce a 'satisfactory' result. Typical conversation is illustrated by the following extract:

HANNAH G   I'm using the compass to make little holes.
CHRISTOPHER A   Can you make a hole?
AMY H   I've never used them before.
DONNA   My arm's come off.

[Louise pulls Donna away to help her.]

DONNA   How shall I stick it back on? How shall I do it?
SL   Try a paper fastener.
HANNAH H   My wings aren't moving properly.
SL   Try re-tying the string.
HANNAH G   I can't make it work.
SL   Does that look like the bird you want? It's a good idea.
HANNAH   What shall I do?
DONNA   I can't do it.
DARREN   I need the string.
AMY H   Is this OK?

The following week I started by giving a short introduction to the lesson, indicating what I thought they should be doing at this stage of the project. I started by suggesting they approached the central resources table and gathered together all the things they required to start work. That was quickly accomplished and I walked around the room looking at pupils' work and gave some words of encouragement where I thought it would help. I could see quite a few 'good' designs and prototypes taking shape, and was conscious of my concerns to judge the models as they took shape more than to judge what was being learnt and how the children were thinking. I indicated that I considered some designs to be 'good'. This was my own opinion and may not have been the same as the child's. The lesson was very busy and the pupils needed a lot of help with skills. I had to spend time explaining techniques and helping pupils use some new materials and equipment. It was during this lesson that it became apparent, by studying the models made so far, that some pupils were having trouble achieving the goal I had set. Amy approached the desk looking very worried:

SL   What's the matter?
AMY S   The wings are going this way instead of that [side to side] not going up and down.
SL   Why do you think it's doing that?
AMY S   If I was to connect them in another way would it be better?
SL   Thats a good idea, go and try it.
SHAUN AND CHRISTOPHER   Where is the string?
SL   Over there on the table.

[Shaun and Christopher go over to the table and find the string.]

HANNAH G   What shall I do, I've finished?
SL   Does it work?
HANNAH G   Well, no not really.
SL   Well, it's a good prototype, work on the design and have another go.
HANNAH G   I know what I'll do. I'll go and do it and I'll show you later.

I believe Amy approached me already aware of her 'mistake'. She was seeking my permission to use more materials to have another go at achieving a better outcome. Another had quickly made a 'good' model but he wanted to see if he could do something different. He started to make another model using similar ideas to two other boys, and used their design to build an improved model for himself. One child who is very quiet and often absent from school, and appears to lack background experiences in using materials and equipment, was not progressing well. I gave her some hints hoping that she would change the basic shape of her bird. These hints fell on stony ground. A short time later she teamed up with another girl who was also having problems. Eventually they achieved an outcome between them.

Shortly before the end of the lesson we stopped and gathered together to evaluate the models that had been made so far. The children studied each others' models and it was the general opinion that one was the 'best'. It seemed to do everything it should. There was a general buzz of conversation. I heard several children discussing new ideas with friends. It seemed to me that they were looking forward to the next lesson.

In the fifth lesson the pupils collected together all the equipment and materials, placed them on the central table, and began working with enthusiasm and apparent confidence. I reminded them that there was one more lesson on the task. My impressions of their progress were noted in my diary:

> Claire approached me, she had taken her first model home, which had been completed. I was disappointed as it was quite good and I thought she would not have time to make another model. She had some new ideas and started work again on a different version.
>
> Donna had lost her wings. She continued throughout the lesson to ask questions and tell tales on other pupils. She had achieved very little since the class had started making. It would appear that Donna needs more help and guidance than some of the other pupils. She seemed to me to be the least able to formulate her own ideas.
>
> There were two other pupils with similar problems. These two did not continually seek my attention but sat quietly, apparently working. It was not until we looked as a group at all the completed models that I realized how inadequate these pupils' attempts at the task were.
>
> Christopher A was less confident producing his second model, he needed a lot of help to execute his design. James had spent a few weeks doing anything but making his model. He is extremely good at doing jobs and using this as an excuse not to do his work. At the start of the lesson I had to remind him to set to work as there was only one week left.

By the final lesson the models were virtually all completed. None of the children appeared to have worked from their original design in making their model. Their designs looked nothing like their finished product, raising questions for me about the expectation to 'design' in technology.

They nearly all produced 'a bird with moveable wings'. There had been only two failures. Each model looked different and showed that my change in teaching style had influenced the children's products to this extent.

By the end of the six lessons I felt the children had changed. At first their reactions seemed to indicate lack of confidence in their situation, the virtual silence a remnant of my previous expectations of class behaviour. The change had been gradual. Confidence appeared to me to develop slowly and by week five an air of knowing what they were doing prevailed for most pupils. In fact in Lesson 5 they started without me, and failed to realize I had appeared as they were working so attentively.

I believe that my new teaching style had encouraged 'independent thinking' and many of the children had enjoyed and attempted problem solving for themselves. The start of problem solving in Lesson 1 and 2 had been hesitant but that soon improved and most children created their own variants on the theme of the task.

The new method I employed had improved my sense of relationship with the pupils. There is no evidence for this except my own feelings. The pupils seemed to appreciate their freedom of choice with materials and what they were making. I had really enjoyed allowing the pupils some leeway in choice.

I had feared materials might be wasted, but the pupils were very economical. There was some evidence that children preferred the new method of working. It was definitely better from my point of view to be working with pupils displaying enthusiasm and enjoyment of their own work. The evidence indicated to me that children preferred to be active investigators rather than passive learners. The lessons had a much better 'feel' to them than they had when I expected the children to be passive. The element of invention had given the children opportunities to invest their work with at least some 'personal meaning'. They were able to make their own modifications as the work progressed. The evidence could be seen in their work books, where I had asked them to draw their original designs, and at the end I had asked them to draw their completed models. In discussions it was obvious that they found the work pleasurable and worthwhile. They were anxious to show the work to their class teacher and science teacher as it related to the work they were doing with them.

The pupils had begun to evaluate their work and make comparisons with their peers. They could discuss whose models were 'good' and the criticism which they offered was constructive. They had made helpful suggestions to each other on how the models could be made to work more successfully. I felt that from their general reactions to the work and the interest it generated the children had enjoyed extending and broadening their technological experience. The ideas that some put forward about what they would like to make next is evidence that indicates they appreciated the opportunity, and would be capable of generating their own topics. But I could not be complacent. These 'new' lessons still left many questions about how best to provide for different capabilities, to maximize learning for every child, and to accommodate individual ideas.

FURTHER READING

Bentley, M. (1990) *Primary Design and Technology in Practice*, London: Longman.
Department for Education (DfE) (1995) *Design and Technology in the National Curriculum*, London: HMSO.
Design Council (1987) *Design and Primary Education*, London: The Design Council.
Dodd, T. (1978) *Design and Technology in the School Curriculum*, London: Hodder & Stoughton.
Dunn, S. (1990) *Design Technology: Children's Engineering*, Lewes: Falmer Press.
Hopkins, D. (1985) *A Teacher's Guide to Classroom Research*, Milton Keynes: Open University Press.
Idle, I.K. (1991) *Hands on Technology*, Cheltenham: Stanley Thornes.

Jeffrey, J.R. (1991) 'Design Methods' in Shaw, K. (ed.) *Teaching Design and Technology: Perspectives 43*, Exeter: School of Education, University of Exeter, pp. 51–65.

Kasick, O. (1991) *Resources and Projects*, London: Thomas Nelson and Sons.

Marshall, A.R. (1974) *School Technology in Acton*, London: The Universities Press.

Shaw, D.M. (1978) *Design Education for the Middle Years*, London: Hodder & Stoughton.

Shaw, K. (1991) *Teaching Design Technology*, Exeter: School of Education, University of Exeter.

Williams, P.H.M. (1985) *Teaching Craft, Design and Technology, Five to Thirteen*, London: Croom Helm.

Williams, P. and Jinks, D. (1985) *Design and Technology 5–12*, Lewes: Falmer Press.

Wisely, C. (1988) *Projects for Primary Technology*, London: Foulsham.

Chapter 9

Mental images and design drawing

Andrew McCandlish

Two important conceptual stages in designing and making technological products are 'imaging' and 'realizing'. Drawing is one of the bridges between the two. Different kinds of views are held about drawing in educational and psychological circles. One view is that drawing affords a convenient window into the child's mind. That underlies certain methods of psychological research and theories of cognitive development (Lowenfeld and Brittain 1970; Freeman 1980; Gardner 1980). Another view is that drawing is useful in the planning stages of the design and making processes as a means of communicating ideas, intentions and proposals. The Programmes of Study for the national curriculum, and the levels of attainment statements, clearly define drawing as a tool in that process, in association with the 'modelling' of ideas by other means (DfE 1995: 2). I set out to explore something of the nature of imaging and drawing, trying to relate curriculum expectations, classroom experience, and my understanding of children's use of drawing in technological activities.

There appears to be little practical help for the classroom teacher when interpreting children's drawing in the context of national curriculum design and technology, and little guidance about how to teach design drawing to young children. I felt that I needed to develop my understanding of:

- what is a meaningful drawing in these activities;
- how the use of technological drawing can best be introduced and developed;
- how the stage of 'making' is represented in drawings.

I was particularly interested in the issue of the usefulness of drawings in the technological process as a whole. In order to address it I chose first to investigate how children regard their drawings. Some sub-questions I had in mind were:

- are drawings regarded as intrinsically useful, or are they seen as a hurdle to be overcome before the teacher allows the child to proceed with actually making something?

- to whom are drawings useful?
- what are possible starting points for a child embarking on a drawing?
- what might be included in drawings?
- how might I foster the use of meaningful drawings in the design and technology curriculum?

Consideration of these questions in that sequence will form the pattern of this chapter, based on research undertaken with my class, which comprised seventeen boys and ten girls, from Years 3 and 4.

Drawing is used for a variety of purposes in my class. In art children use a variety of graphic media, representationally and non-representationally, to develop skills and explore themes through imaginative and observational studies. In humanities drawing is used to record scenes, illustrate costume, transport, or buildings, and to represent people from different places. In science accurate drawing to record objects is demanded. During breaktimes children draw for their own interest.

There are differences in drawing 'style' which seem to relate to 'audience' as well as to purpose. For display, drawings tend to be carefully proportioned, detailed, and searching for 'realism' with the use of light and shade and the beginnings of perspective. Private drawings regularly feature cartoon-like caricatures. Drawings of places often convey both plan and elevation, in the characteristic style of this age group described by Lowenfeld and Brittain (1970: 157).

Lowenfeld and Brittain (1970: 37–8) suggest that children of this age are at 'preschematic', 'schematic' and 'dawning realism' stages of representational drawing. The first of these is characterized by objects and figures placed somewhat randomly on paper, and of variable scale. In the second there is spontaneity and willingness to discuss the content of drawings, which is derived from a logic of conveying the main features of, and relationships between, objects. A greater interest in detail characterizes the last of the three phases, with a concern to render appearances as literally and true to life as possible.

Incidental and unsolicited comments from the children in my class reflect a wrestling with drawing problems akin to representing the interior of a closed object. There is evidence of struggles with projective representation and perspective (as suggested by Arnheim 1972: 194). A tendency for children at this stage to strive for increasing accuracy, favouring language rather than drawing as a means of self-expression (Gardner 1980: 150) seems to be reflected in a comment by one child in my class: 'In my head the thing looks better, because my hands can't do better than my brain, and my hands can't do as easily as I think and talk.'

In order to investigate how the children regard their drawings within design and technological activities I observed them within a normal learning context: during a problem-solving task, set within a cross-curriculum

Mental images and design drawing 97

Figure 9.1 Examples of the children's design drawings

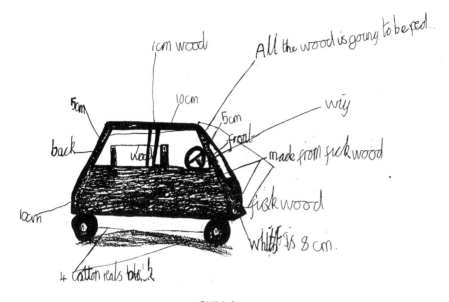

Child A

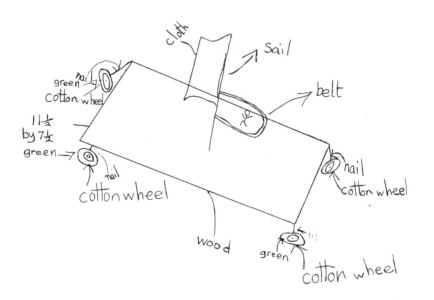

Child C

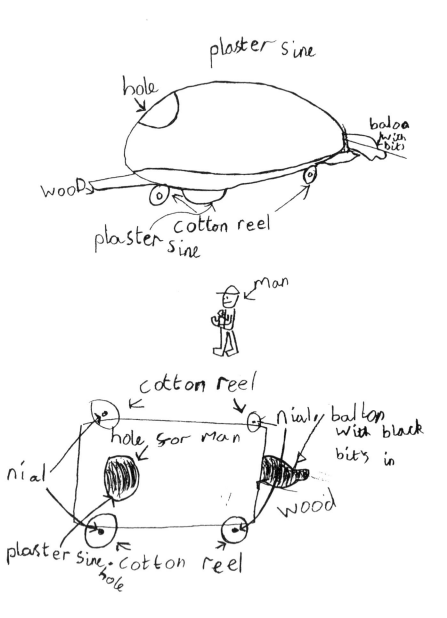

Child F

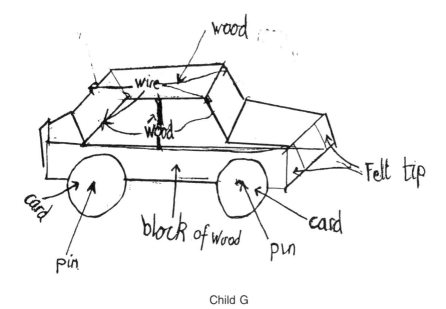

Child G

topic on transport. The task entailed designing and making either a vehicle to carry a standing Lego person for a distance of one metre; or a model of a vehicle which, if enlarged, would suit a family.

The children were asked to produce at least one drawing and then to discuss it with me. They were expected to explain features of their design(s) and the materials they needed for making the vehicle. This enabled me to ask questions relevant to the research within a normal classroom context, avoiding interference with their learning, and reducing the potential for extraordinary behaviour or responses. Some discussions were tape-recorded, then transcribed. The record which follows is a selection of data following the sequence of my research questions.

USES

Recognition of two strategic uses of drawings were identifiable (though others may exist): helping to clarify ideas, and as an *aide-memoire*. The notion that committing ideas to paper entails a developing and refining process seems to underlie several sections of the Programmes of Study of the design and technology curriculum. In practice it might be difficult to demonstrate this, and its relationship with other thought processes is unclear. Children talked about their design drawings by referring specifically to the components and materials which they intended to use, and functional features of their designs, such as drive mechanisms or holding devices for passengers. This can be interpreted as evidence of the translation of their responses to the task into design details. Drawings recorded their proposals, and perhaps provided opportunities to convey the initial design in a disciplined way, by thinking though the various 'needs' of the vehicle. Omissions of detail seemed to become apparent as discussions of drawings projected attention towards the making phase.

AMcC Could you make something without drawing it first?
J Not really, you have to see what you need. It helps you think things through.

Other children recognized that in the evolution of a design ideas became outmoded. This didn't necessarily render drawings superfluous; rather it seems to suggest that these children were confident that their ideas were viable enough at the stage they were recorded. Evolution seemed to have been limited to the addition of detail, because the children largely stuck to the main structures which formed the core of their designs, amending smaller items such as wheels and restraints. Often this was in the light of increased knowledge about available materials and 'better' alternatives. Sometimes it was recognition that initial intentions were not realizable:

K Sometimes you can't do what is on the drawing so you have to change it (during the making phase) to make it work.

I was careful to use the word 'drawing', but used the word 'plan' to respond to three children who themselves referred to their drawings with this term. Plan might imply the attainment of clarity or precision, or set of intentions. This seemed like further evidence that many of the children saw drawing as making a significant contribution to clarification in the whole of the design process.

The use of drawings to assist making is explicit in the Programmes of Study. One possible interpretation of this is as an *aide-mémoire*. The children generally recognized this as a principle, but in practice there were differences in its application. On one hand D was observed with his drawing squarely in front of him (with no prompting from me) using it for ready reference at the making stage. At the other extreme E was not seen using her drawing at all, and claimed not to need it. J said he referred to his drawing before starting work so that he could see what next needed doing. K supported the principle clearly, though comments about his practice showed that he distinguished between circumstances in judging the usefulness of drawings:

K If you get in a bit of a muddle you can find out what to do next.
AMcC Could you make things without drawing?
K Probably, because sometimes you do the drawings of really hard things and sometimes you can make things off the top of your head, you know what you are doing.
AMcC Which is the car?

K   Top of the head job.
AMcC   So did you look at the drawing much?
K   At the beginning I did.
AMcC   Why not later on?
K   Because by then you've got the basic idea.

Other pupils had similar views. G said that if he got stuck he would refer to the drawing which would help him to see 'what to do'. However, he claimed not to have used it much since the details were so simple that he had memorized them. Perhaps this suggests that the simple level of the task and the short timescale between designing and making stages made this use of drawings largely redundant.

AMcC   Which came first, your idea about the shape or the idea about the things you were going to use?
A   The things I was going to use.
AMcC   Why did that come first?
A   I think about how I was going to make it and then drew it.
AMcC   And you think this is a really great design, do you? You couldn't possibly do better?
A   I could put a bit more colour into it.
AMcC   Did you mention colour before? Is it important?
A   Yes.
AMcC   Why?
A   So it will look nice.
AMcC   On the design or on the actual car?
A   On the actual car.
AMcC   Has your drawing been of any use to you?
A   I really looked at it and I learned it off by heart – there's only a couple of bits to do so it was easy.
AMcC   Is there any point in doing a drawing?
A   Yes – it tells you what to do. If you have a drawing it stops you from going wrong.
AMcC   Could the drawing be wrong?
A   Yes. If you get the wheels wrong it won't work.
AMcC   When does that become obvious?
A   When you've made the model.
AMcC   Could you make the model without it?
A   No. It would turn out a complete disaster – you might keep changing your mind.

Another possible interpretation might be that drawings exemplify general outlines: for example, a vehicle being represented as a box with four wheels. Making would follow, with elaboration on that theme taking place at that stage. Child I's work (see diary extract which follows) was seen this way, whereas other children's drawings contain more detailed proposals. Whether there is a link, and what kind of link there may be, between cognitive development, motivation, attitudes, and the complexity of the drawing, which may then increase in value as an *aide mémoire*, is a matter which I will need to explore, especially in the light of I's work, recorded in my diary:

> I drew an outline which was vaguely car shaped. To make it he went to the wood box and assembled a huge pile of assorted shapes and of the largest pieces. He got two inch nails and three quarter inch panel pins and proceeded to knock them in, making a two inch row of pins and nails, towards one end, across a two foot long, six inches wide, half inch thick plank. He then extracted the nails using pincers and hammered four panel pins into the narrow edges, each towards the extremities of the length of the wood, as if to make axles on a flat-bed chassis. He then fitted cotton reels to these. Next, he placed differently sized and shaped offcuts with their narrowest edges lying on this chassis as if to represent sides. No attempt was made to shape these offcuts or to secure them to the chassis. At this point imbalances of the offcuts and insufficient penetration of the panel pins caused a fairly comprehensive collapse. At no point was any reference made to his drawing. Advisory discussion initiated by me took place: the following points emerged. He:

- expressed little confidence in his drawing ability;
- saw the drawn design and the making of the vehicle as being connected, but only insofar as they both were concerned with a vehicle;
- said the drawing had no planning function;
- spent a lot of time looking around at others' models – the chassis design was derived from that of a named peer (whose chassis was considerably smaller);
- had changed his mind 'quite a lot' whilst thinking about his vehicle, and he still had no clear idea about what it would look like – he just wanted to get on and make a vehicle quickly.

None of the drawings contained any note of the sequence or priorities for the making phase, and no difficulties in sequencing were observed or reported during that phase. I was led to conjecture that this aspect is contained in an internalized 'hidden plan'. If that exists, together with recall of the details of the drawing, then a significant number of the children would be unlikely to refer to their drawings in this phase. The relationship between drawing and the use of modelling with materials and components would need further investigation. For the moment there were other immediate considerations, such as the likeness of the drawing to the model produced, and the evolution of the design at the making stage.

In all but one case it was possible to observe likenesses in features such as overall shape, position of wheels, and method of restraining the Lego figure (in that specific task). The conclusion can, therefore, be drawn that there is a relationship between the drawings produced and their use in 'imaging' for the making process and the final product.

During both the drawing and making stages children discovered that there were reasons for modifying their designs from those drawn. One instance of this occurred during the interview with pupil B:

AMcC Where does the balloon go when it is blown up?
B Ah!

He had failed to allow for the additional displacement. Although he did not amend his drawing, he proceeded with flexibility to allow for this during the making phase. Other children expected evolution of the design during the making phase:

AMcC Do you think that it is going to turn out exactly as you've actually seen it, or as you've actually drawn it?
G It might be, I don't know.
AMcC Why might it not?
G Because it might not have the things that I want it to have. [reference to availability of components]
AMcC to K Do you think it's going to actually turn out like that?
K It might. It might go wrong and I might ... do something different and something might go wrong on it, like I can't get the stuff – I can get most of it in the classroom.

It might seem that drawing could be regarded by some children as a hurdle to be jumped before the making phase since they could possibly make a vehicle that satisfied the task without making a drawing. The evidence suggests, though, that drawing and discussion offer opportunity for deeper thought and improved design proposals, but leaving room for further improvements in the making phase.

USERS

I had anticipated that in addition to the individual child as 'user' of the drawings, two other users or groups of users may be significant: other children, and the teacher. None of the children mentioned the possibility of other children using their drawings. Given that the task was to be tackled individually, the possibility of peers using their drawings is remote. Ownership of design work may be regarded as highly personal, and that could be reinforced by the way the task was set. This does seem to be problematic when compared to practices of some design-team activities.

There was no evidence that the child-designer might regard the teacher as a possible user. There are grounds, however, for believing that the children regarded the teacher as evaluator of the ideas conveyed by the drawings:

AMcC Why do you think I asked you to make a drawing?
E To see if we change our ideas. To see if our ideas work.
AMcC Is [the drawing] any use to me?
G Yes if you wanted to know if [the design] is possible – or give tips.
AMcC Who needs the idea [as shown by the drawing]?
J You [the teacher] do.
AMcC Like a check-up.
J [affirmative nod]

Tips could be advice as to techniques for using tools and materials. During the drawing phase children often sought assurances as to the appropriateness and quality of their work – a common feature in a primary classroom.

STARTING POINTS

Given that some sort of mental process precedes and accompanies drawing, drawings may represent some priorities in design considerations which occur in response to tasks. The lack of research in this field reflects the difficulty of entering another person's mind. Despite this, if opportunity is given for children to reveal what is uppermost in their mind at the pre-drawing stage, some understanding of their drawing might be gleaned. From the data I gathered an important starting point seemed to be the visualization of the product which would result from fulfilling the task. Some children claimed that they had clear mental visions of their end-product:

> I got a picture in my head. It [the vehicle] was going to have a sail and a hole where the Lego man was going to stand, and a piece of wire attached to the sail and around the man's waist. The idea is for you to blow the sail and it would sail away ... like a land yacht.
>
> (C, who went on to describe a land yacht)

The use of the past tense to describe the mental image is a result of the timing of our discussions (it would be particularly difficult to capture the imagining process as it occurs):

D The idea of a sort of golf car came into my mind.
B I imagined a kind of picture.
A Little mini cars I've got – that's how I thought of this car. It came up in my head so I just drawed it down.

These statements seem to suggest that an image seen in the mind's eye substantially influenced the drawing, since none of the objects described were physically visible at the time the task was set and the drawings were made. The source of such images was largely (and unsurprisingly) the child's personal 'experience', which was not always conveyed in the drawings:

AMcC   What came into your head straight away ... was it this design?
H   No.
AMcC   Could you describe the one that came into your head?
H   It was mostly like my own car.
F   I'm thinking about a van.
AMcC   Have you a picture of it in your head ... what is it like?
F   Yes, in my head the thing looks better.

This problem of translating image to task was also evident in responses where children seemed to give priority to the materials that they thought would be suitable and available to the production of 'their vehicle'. The constraint in the task limiting materials to those available in the classroom was crucial:

AMcC   When I gave you this problem how did you see your solution?
D   I thought to make it out of wood.
AMcC   How did you come to think of this idea?
E   A piece of wood with a hole for a Legoman.
F   I thought about the materials I would use because that's more important than the design.
AMcC   Which came first, your idea about the shape or the idea about the materials you were going to use?
A   The things I was going to use.

Which starting points came first was difficult to establish from my data. It would seem various possibilities included:

- whilst hearing the task requirements the child's mind interacts freely and creatively with remembered experiences;
- priority of a mental vision is important, but there is scope for material considerations to co-exist with that priority;
- design possibilities are modified if and when restrictions are announced;
- designs are modified by the process of translating mental image into drawn image.

To press these questions would require further research. The interactions of vision, drawing, and awareness of material considerations strongly suggest that drawing serves a purpose by affording opportunity for children to record their first ideas for solutions to tasks, and thereby explore task boundaries. At such an early stage in the design process this could include the facility of thinking with a pencil (Kimbell 1982: 45) – rapid drawing where visual expression keeps pace with thoughts.

But it also suggests that this mediation creates its own difficulties of a mismatch between ideas and realization.

CONTENT AND ELEMENTS

The sort of details that a drawing contains may show how meaningful it is to the child, as a part of the mediation between ideas and product. For both the child and the teacher the detail may reveal, or contribute to an understanding of:

- a perception of what is deemed important in planning a product;
- the development of a coded style of drawing peculiar to technology as part of planning and preparation;
- the interest of the child in the planning process;
- the level of awareness of design considerations;
- the strategic uses of the drawing towards successful products.

Given the variety of potential content of drawings, the range of modes of visual representation available, and the choices of visual elements such as shape and colour, I decided to consider how children used colour, shape, dimensions and materials in their drawings.

Few of the children's drawings were coloured. They were mainly done with graphite pencil. Within the context of drawing in the classroom this was abnormal. In all other curriculum areas coloured drawings are more common than black and white, and I usually clearly specify this latter condition when it is required, which I had not done on this occasion. Previous relevant experiences which may have conditioned this response are difficult to assess. Children may well have seen house plans and simple technical drawings, for example to explain the workings of a machine; but they all have had experience of working from Technic Lego cards – which are in colour.

Some interview conversations raised the issue of colour. Child A said that colour was important, not to beautify the design but rather on the actual model. F referred only in conversation to the colour of the vehicle which inspired his design. On his drawing the exhaust was heavily blacked. When the question of colour was raised with G he initially objected: 'I think it might spoil it (the model) if you put colour on it.' It transpired that this could be attributed to his limited knowledge of colouring agents suited to his model. H felt details of colour were unimportant, but one week later she revised her plan, specifying a colour scheme. This may have been as a response to what she felt was a hint by the teacher. In the making she seemed to lose sight of the colour scheme – only the wheels were true to the colour specified on the revised drawing. K explained that his model would be blue, because his family liked blue. This was only revealed in discussion, not in the drawing.

From this data it appears that although the children may have considered the question of colour in their minds they did not always commit it to the drawing. It would need further specific investigation to determine the reasons for this. Possible reasons might include:

- colour would obscure other necessary detail on the drawing;
- other details, such as which materials to use, are uppermost in the mind and the child loses sight of the detail of colour whilst wrestling with more pressing issues;
- limited technical knowledge of appropriate colouring agents make colour a detail that the child would prefer not to raise deliberately;
- the child has established no colour scheme for this context (cf. Lowenfeld and Brittain 1970: 164).

Shape would be expected to be of fundamental importance. By this age representational drawings are familiar currency (cf. Gardner 1980: 148), and shape is perhaps the most important element in representation. In all cases this was shape delineated by outline. Only child I's drawing bore little literal resemblance to the shape of the model he began making. This may well have been a reflection of cognitive factors, in the sense that he was not (yet) able to draw shapes, in combination, sufficiently well to record the intended design of his vehicle. Such accurate translating was certainly important to others, who seemed to offer clear indications of how shape was derived in their drawing:

A   It came into my head so I just drawed it down.
AMcC   Exactly the same? No differences?
A   No.

Here the child claims a literal translation of image to drawing – an attitude akin to that observed in children of similar age which was, perhaps unsympathetically, described as 'a pedantic preoccupation with the photographic aspect of drawings' (Gardner 1980: 148). This may be true of B:

B   I imagined a kind of picture. A wooden platform with a rectangle of wood around – a tight rectangle to put the Lego man in and a balloon at the back.

In the particular code of technical drawing size is commonly represented by either including dimension lines or by showing a scale ratio, for example. The children were aware of scale, recognizing toy representations of vehicles, for instance, and appropriately responding to instructions to draw full size or larger than life in Art. The drawings in this task contained only occasional indications of the intended sizes of their vehicles. However, there was a ready response from children asked about the size of components represented in their drawing:

AMcC How big is this base (going to be)?
H Fifteen centimetres by eight ... seven centimetres.
AMcC What made you decide on that size?
H Well I got a ruler out and I thought five by ten centimetres would be too small, and ten times twenty centimetres would be too big – so I picked a number in between that.

K explained that the drawing could be measured to give an idea of size. His drawing happened to be about full size, but since it was essentially a side elevation, the representation of how 'fat' the model was to be was not readily resolved, as he had not thought about that.

One possible interpretation is that if a child is drawing 'literally' from what they see in their mind's eye, some idea of scale is necessarily envisaged but not always expressed by conversion into dimensions for the model. Another is that they record the minimum of information necessary to convey an idea.

The size of the drawing might be influenced by the size of the paper, which itself may be interpreted by the child as an indication of the possible size of the model expected by the teacher. The children may believe that they have given a fair description of intentions by showing details of shape alone. At this stage of their designing experiences the use of measurement conventions among these children was very limited, and the study alerted me to the need to address this matter urgently. Further, since no reference to formal units of measurement were made when setting the task, the children could expect flexibility of size at the planning stage. Size might also be seen in terms of non-standard measures (various parts of the Mathematics curriculum give experience of such), and estimation of size might well be based on the size of component materials provided in specific sizes – such as cotton reels.

There is another possible reason for the sparseness of formal sizing in the drawings if, as here, the children make their own model. Details of dimensions seemed to be present, but largely in the designer's mind. Data which is not to be visible in the finished model could be given lower priority by children who do not understanding the purposes of technical drawing conventions, or do not need to convey the information to others.

A major consideration in the children's thinking at the drawing stage appeared to be determination of the materials from which the model would be made. All the children in the group showed awareness of materials. Alongside shape, detailed indication of materials appeared on all drawings except I's. The importance of materials in generating the ideas emerged in interviews:

AMcC   How did you set about your design?
F   I thought about the materials I would use because that's more important than the design.
H   I found out I couldn't make it like my car because I hadn't got the right things to make it.
AMcC   Tell me about your idea.
J   I thought of this kind of piece of wood with wheels on.
AMcC   Why wood?
J   Because it doesn't break easily.

FOSTERING USAGE

Drawings in technology might take the forms of thinking with a pencil and of formal working drawings (Kimbell 1982: 45–6). The degree of detail in the drawings by children in this study puts them somewhere between the two. It seemed that the drawing phase was not regarded as a mere hurdle to be overcome before the making phase. Drawing offered opportunities to think or aided thinking. Ideas were clarified and evolved during the process of drawing, but not all is shown in the drawing itself. I needed to seek out internalized information in order to gain a fuller picture of intentions – and then evaluate how meaningful the drawing was to each child.

As an *aide memoire* a drawing may be a trigger for recalling internalized information. As a testing ground for ideas it allowed fluidity for a design to evolve during both the drawing and the making phase.

The majority of children saw themselves as being prime users of their drawings. I used the drawings too, but as a means of 'seeing' ideas and proposals, in order to monitor the children's responses, check the feasibility of their plans, and advise where necessary. In this process of teaching I recognized the need to encourage the children to think of design drawing for other 'users'. If the notion of drawing as a communication tool is worth pursuing children might work collaboratively, exchange drawings, and make received or collective designs. A benefit would be the need to consider the quantity and quality of detail, including awareness of dimensions and assembly sequences. A pre-requisite in such work would be self-confidence and mutual respect, and appreciation of the nature of teamwork.

The drawings produced were essentially freehand sketches, using one viewpoint: a familiar code to the children. Other approaches to technological drawing include the use of isometric projections, and plan/evaluation/section drawings. The concepts for these may not be developed among lower juniors (cf. Lowenfeld and Brittain 1970: 148; Gardner 1980: 216–20). The use of annotated sketches, I decided, deserves serious consideration as a possible forerunner to these more formal modes of drawing. Leonardo Da Vinci used sketches for designs (Goldscheider 1948: 18)! But it was clear that more thought would need to be put into my teaching strategies for developing their quality and their use in ways which were 'authentic' and purposeful in the design and making process.

Since during the lower junior years children's general drawing is very much subject to evolution attributable to the process of psychological maturation, learning experiences and teaching strategies will need to approach this aspect of technology with care. My initial investigations showed that the strategies which I adopt will need to take account of the children's perceptions of design drawing, and ensure that it is indeed developed as a purposeful medium in their terms.

FURTHER READING

Arnheim, R. (1972) *Art and Visual Perception*, London: Faber & Faber.
Department for Education (DfE) (1995) *Design and Technology in the National Curriculum*, London: HMSO.
Freeman, N. (1980) *Strategies of Representation in Young Children*, London: Academic Press.
Gardner, H. (1980) *Artful Scribbles: The Significance of Children's Drawings*, London: Jill Norman.
Goldscheider, L. (1948) *Leonardo Da Vinci The Artist*, London: Phaidon Press.
Kimbell, R. (1982) *Design Education – The Foundation Years*, London: Routledge & Kegan Paul.
Lowenfeld, V. and Brittain, W.L. (1970) *Creative and Mental Growth*, London: Macmillan.

Chapter 10

Working together

Annette McMylor

I have been interested in design and technology throughout my teaching career but still do not feel confident about teaching it. What I have learned about the subject has been collected on INSET courses and from teachers with more specialist knowledge than I have. However, I now feel that the best way to proceed towards the development of my teaching is to embark upon a topic and learn with the children. Only in this way do I feel the children will get the opportunity to broaden their experiences and increase their technological capability.

I have no technology equipment in the classroom. Whatever tools and materials I need have to be brought in for the duration of a project. There is a designated design and technology room in the school, equipped as a more traditional, specialist workshop. I see it as an inappropriate area for my class to use, much preferring to work in the classroom base, with general facilities. However, the floor is carpeted.

Up to now I have introduced technology projects and tasks and left the children to find their own solutions. I have become increasingly aware that I need to provide a more solid base of skills and knowledge, and a wider variety of experiences which can be applied to a range of problems. Previously technology work has also been organized so that children worked individually, or at most in pairs, using card as the main material, to make small scale products.

Until recently this was not unusual in the school. Technology was viewed by staff with some trepidation and had been neglected. There was no coordinator and no policy. It appeared that little time or money had been allocated to the subject. INSET had been somewhat disjointed. Although helpful with some 'ideas', it did not provide the boost in confidence we had hoped for. In particular I was still left with concerns about time allocation, lack of availability of materials and other resources, and the inexperience of children who entered my class. There was no evidence of skills the children had already encountered, so we tended to 'start again' when they transferred into this 8–12 middle (deemed primary) school.

Although the school has been very 'subject based' in the way the curriculum is organized, I had previously worked in a cross-curricular way, through project work, characteristic of many primary schools. I have always had a policy of encouraging children to work independently, without too much reliance on, or interference from, me. With the advent of a new regime at school the opportunity arose to work in that way, with group work as a central organizing feature. I was conscious, though, that eight-year olds have to learn how to get on together, how to work cooperatively, and that one cannot assume that good group work just happens.

Most of the children in the class came from the same local first school, and had been together since they started school. When they came to me as their Year 4 teacher I thought this would be to my advantage, that they knew each other and would be used to working together. I soon discovered that in fact they were very quarrelsome and relied heavily upon adult intervention to sort out difficulties between them. It is a class of twenty two mixed ability children. At the start of our new project – the Egyptians – I did not feel that they were responding to the group work organization. In particular the aims of considering others and valuing their own achievements seemed to be at risk.

Many teachers I have seen organize their class in groups, and doing that successfully is something I am still striving for. However, I am aware that that is not easy. Stanford, for example, points out that 'most classes never develop into groups. They remain a collection of individuals, lacking the attitudes and skills needed to work together effectively' (Stanford 1977: 3).

For the children, thrust into peer groups, the task is not easy either. Many of them take a long time to adjust to their peers, and I am very conscious of some children in the class still trying to come to terms with others. Downey and Kelly highlight what I feel is fundamental to the situation: 'Friendship patterns among young children are not very stable and a group working together amicably one week will be at odds the next because of some apparently trivial disagreement that looms large in the children's minds' (Downey and Kelly 1979: 15).

I was certainly finding this to be the case, and was trying to get the children to be tolerant of each other, to think more about how what they do affects other individuals and the whole atmosphere of the classroom. When we began the project I knew that they had a long way to go in learning how to respect the opinions of others, and in learning how to get on together. The vast proportion of my day was taken up with sorting out upsets and arguments. I decided that I needed to investigate exactly what the children were doing when they were working in groups, particularly during technological activities, where the need for cooperation and collective decision making is often essential.

I kept my observations in focus to ensure that I looked for specific data to illuminate the ways they responded to group work. I looked for differences between the groups to see if this affected the end results of their activities. To this end I considered what the children were actually doing while 'working together', i.e. how they talked to each other, and the dynamics of each group. I chose to keep out of the way as far as possible to ensure that their behaviour was not altered by my presence. I expected this to be difficult. I have often thought that the only way to observe a class would be for them not to know they were being observed. That could not happen in this case. The data was gathered through my own observations during lessons for which I was responsible, and recorded as field notes. I could see many problems with this method of recording data. Even with the groups working 'independently' I was still in a teaching role. I also knew that there would be events that I would miss, and therefore gaps in the data when I came to analyse it. I drafted a check-list of specific things to look for and record:

- who was leading the group?
- were boys and girls working together?
- how were the less academically able children coping with the task?
- when did arguments arise and how were they resolved?
- did the children remain on task?

I also used a camera to record specific moments, and audio-recording to try to capture the language of interaction of groups, even when I was not observing the events of each one. The tapes provided a bank of information from which I could reinforce observations and illustrate particular characteristics of group work. In addition to this eavesdropping I also recorded discussions with the children away from the task, to establish how they saw their experience of working in groups. To supplement these records I asked the children to keep their own notes in the form of a log of events, and to present their designs at the end of the planning task. At the conclusion of the whole project they were asked to present their model to the class and report orally on what they had done. A tape-recording of the presentations was made, though this provided limited data relevant to the ways the groups had worked together.

The Ancient Egypt project lasted the whole of the Spring term. At the time of introducing the design and technology task – to design and make a working model of a *shaduf* – we were considering farming methods. I anticipated a burst of initial enthusiasm and did not think the model would take long to build. I envisaged there would be more problems when it came to making the *shaduf* work. I knew from the previous term which children would be likely to squabble, but I allowed free choice of working partners, with the guidance that they should think carefully about who they could work well with. Five pairs and three groups of three emerged, based entirely on prior friendships. They were all single sex, and only one which I regarded as of 'mixed academic ability'.

Following other activities relating to farming methods in the Nile region, and the need to store and distribute water to the land, I used a series of pictures to illustrate how the latter was achieved. The pairs/groups were asked to work out how the *shaduf* functioned, and to record their explanations on tape. The task of designing and making a working model which would help them to water plants, in a situation like that of the Egyptian farmer, then began. The children took 'shaduf' to mean the basic shape and principle, which could be altered or adapted in any way they chose. I asked them to try to think of several designs for possible solutions to the irrigation problem. Their 'log' was used to record their ideas. We began work on January 26, and the following account is an outline of the way the work progressed, based on the data collected.

As I moved around the classroom the pairs/groups began to discuss how they were going to make their model, but initially it was individuals who asked me to help. I became aware of what I saw as strong clashes of personality, and realized that the children were relying on me to sort out their differences. I took the line that I would not arbitrate, since I intended that they should learn to work together and to reach agreements between themselves. The group notebooks began to show that within the pairs/groups they individually drew versions of the same idea. An exception to this was Kathryn's group, whose members couldn't agree upon a design. They each produced their own, from which they said they would choose the 'best'. I spent a long time listening to this group trying to reach a decision.

More discussion occurred in other groups when they began to consider the materials they would need to make their models. Rob's group in particular spent time deciding who could bring what from home. They talked about using cardboard tube, a fishing reel to wind the bucket up, and guttering to transport the water. At the end of the lesson they recorded their thoughts on what they had done.

The decision I made to stand back and observe how the children worked together – rather than intervene in the way they did so – left me with the view that the majority worked as individuals, without agreements as to who should record ideas 'for the group', and, indeed, not much discussion about which ideas should be recorded. At the start of the second session four previously absent children rejoined the class, and others were away. Gaps were filled; some pairs extended to groups of three; and the membership of some groups altered (though this turned out to be temporary as they shifted back to rejoin original partners). Materials, tools and equipment were made available to start the modelling.

Kathryn's group continued to argue over the drawing, and I eventually spent a long time involved in the discussion, trying to steer the girls towards a decision. They could not see how their drawings were similar. They were worried that if they chose their own drawing as the 'best' it would mean the designer would have to do all the work making the model. Their discussion, however, was reasoned, trying to think of ways of choosing, such as getting an independent person to pick, or to pick names from a hat. I left them to sort this out.

Rob's group had three similar drawings, from which they chose Chris's, because it 'have got more detail and we now it well work' (group log). Nadine had a clear idea of how her model was going to work and had already decided how to anchor it. Her partner was not so sure. Meanwhile Kathryn's group was on the verge of each making their own model. They were aware that this was a difficult time. Lauren wrote in the log: 'The hard bit was when we had to decide what picture to make [into the working model].'

In another corner of the room Cheryl had a very clear idea of her own design but it was complicated. Her partner wanted them each to produce a prototype and then see whose was best. Cheryl wrote: 'Suzanne and I couldn't decide which one to make. Miss McMylor came and said we can make two shadufs so we made Suzanne's first because she said it would take more time' (log). Suzanne wrote a similar note. Although they were 'working together' they kept in mind their 'own' models, eventually reaching something of a compromise. Suzanne added: 'Me and Cheryl made up our minds to do her shaduf after that I said to her "in that shaduf we can mack it a bit of main in it"' (log).

Stevie, who was absent when the project began, did not like any of the designs his group had produced and so set to work to come up with something of his own. His drawing was much more detailed than the others', and they agreed it was better, with David pointing out that 'it's likely to work'.

The 'making' stage began, and initial models were completed quickly. It was obvious that some of them would not stand, and discussion about how to make them stand up ensued. I hoped that modifications would follow, but that they would be instigated by the groups' recognition of the problem and attempts to solve it. As the making proceeded I judged that the groups were mostly deploying people wisely to do various tasks towards building 'joint' models. There were some who seemed very busy trying to do many things, and some who seemed happy doing very little, but there were no signs of disagreement.

At this point I took stock of the way in which I thought things were developing, taking account of what I had observed and what the children had said and written. I felt I was losing sight of the initial focus – how the children worked together – and was becoming more concerned with the quality of what was being produced. The children were working very quickly and their results were to my mind tatty, with much evidence of the use of string and Sellotape.

Tape-recording conversations was proving technically difficult. Keeping the group logbooks was not very successful because nobody really wanted to do it. The children who were most capable of writing for their group were not keen to do so because they saw it as 'having to do all the work'. Agreeing what to record was also a problem. Kathryn's group passed their log book round to collect individual reports of what had gone on. Cheryl and Suzanne wrote about the same events from their individual points of view. Other notebooks were almost empty. I was dependent largely on my own observations and impressions, and these were sketchy.

It seemed that difficulties in working relationships arose more frequently in the larger groups than in the pairs. That, in a sense, might have been anticipated since more individual ideas and views have to be accommodated and negotiated. But the groups were self-choice. I began to realize the complexity of the issues I had begun to explore – what if working groups were chosen by me? What if they were larger? How much time do children need to work out procedures for decision making? What assumptions about ownership, contribution, reward, self-esteem, and so on, are being made?

By the beginning of February the models were beginning to take shape and dissatisfaction within the pairs/groups was growing. Membership of the groups was becoming more fluid, with children drifting in and out of different ones. Leon's group kept changing their minds and were arguing a lot. They seemed to have no clear idea of what direction they were pursuing, and seemed to be just sawing bits of wood with no apparent aim in mind. Leon left the group to sulk in a chair. Charles took over as organizer and tried to stir them into action. George's partner was absent and he was experiencing difficulties because 'it was hard to know what to do next'. Later he wrote: 'It was better and easear with Oscar here because he knows more about the modle' (log book). Rob kept complaining that his group was getting nothing done.

By the following session David had been looking at a picture of a *shaduf* in a reference book. He talked with me about it, and told me his ideas for making the model swing round. However, he would not convey these ideas to his group. Stevie did not want to consider any change to their model. The others left him to think about that, and after a lot of talking they voted on whether they should cut it down in size to give it more stability. Stevie took this badly and was tearful for a long time.

In some groups members were helping each other with skills such as sawing, and finding materials. I was pleased to see signs of cooperation. Some were showing signs of becoming more familiar with the demands of negotiation which were part of being in a group. In the middle of one argument Kathryn said: 'It was only a suggestion.' This was a major advance for a group which had spent as much time arguing as building. Still Stevie's group was in a state of collapse. Rob's group was wasting a lot of time, and making their model smaller was using up more materials. My observations were a little encouraging, but not much. I wanted to know how the children themselves viewed events.

I thought, on the basis of these impressions, that the experience of group work had been largely a disaster for the children. Interviews with them after the event showed that they saw things very differently. Most said that they would rather work in a pair or group than work alone, though some expressed a preference for working on their own or with a single partner. The reasons given for their preferences also provided some surprises:

ROB   It's better to have four than two. You make it quicker, you can do team work.
STEVIE   You can have a team effort.
CHERYL   I think working on my own is best. I don't like working in groups. (However she would like to work with Lindsay) because she doesn't argue too much.
SUZANNE   It's harder working with other people. You argue more.
TRICIA   It'll be an easier decision to just decide between two.
LOUISE [referring to working with Nadine]:   With loads of people we keep getting confused and that, and squabble.
NEVILLE [who preferred to work alone]:   People can't boss me around when it comes to sticking the parts on.
KATHRYN [saying she would work with boys]:   You don't fight with them coz they're different and they get along better than [with?] me.
STEVIE [with a sense of reconciliation]:   We've had our ups and downs but we still got on as a group and worked together.

In the final sessions of the project the children were engaged in tidying up their models and adding finishing touches. They reviewed the finished products in a classroom presentation, after writing an evaluation and preparing an oral report based on it. All the groups were generally pleased with their work, but said nothing about how they had worked together in their groups.

When I had first met them I had entertained a grand notion that they would adapt readily into working in small, self-chosen groups, that they would get on well together, and be capable of reaching agreements about their tasks in an atmosphere of calmness and cooperation. Events showed that classroom reality was far from this idyll. I realized that for the class to work successfully when organized in groups it would need practice on the part of the children, and guidance from me. At the extremes there were times when I had to intervene to restore calm and pacify arguments which were showing no sign of being resolved by the children themselves. Early on it was tempting to think that group organization was more trouble than it was worth. I might well have decided that these children were better provided for working on their own, towards end products that were solely the work of each one. My dilemmas are not yet over: the models are still on display in the school library and deciding who takes them home could rekindle some disputes. However, in the light of the opportunity to watch the class in action and discuss groupwork with them I have reached a number of 'awareness points'. These could each form part of a basis for achieving my aim to encourage successful group work. They include:

- recognition that my impressions may not be entirely shared by the children;
- the need for a more detailed analysis of the aims/benefits of working in groups;
- acknowledgement of the need to devise strategies for careful nurturing of group work;
- identification of the need for children to appreciate the purpose and the processes of negotiation and cooperation;
- a need to understand what is regarded as trivial, and what is of major importance, in the eyes of the children.

REFERENCES

Downey, M. and Kelly, A.V. (1979) *Theory and Practice of Education*, London: Harper & Row.

Stanford, G. (1977) *Developing Effective Classroom Groups – A Practical Guide for Teachers*, New York: Hart.

Chapter 11

Early years children, designers, and partner choice

Kevin O'Grady

Internal staff reorganization gave me responsibility for an infant class of twenty-seven Year 1 and 2 children. This was my first experience of teaching such a young group. I was also made responsible for administering National Curriculum Key Stage 1 assessment to Year 2 – another personal 'first'! In addition to my duties as deputy headteacher and as an extension to my school role as technology coordinator I was about to embark upon my first full year as teacher consultant for a cluster of six village primary schools, contributory to one local high school. An in-service design and technology course provided a chance to focus upon early years classroom organization and practice in delivering technology, enhancing my own skills. These are the 'local' factors that were predominantly to influence the direction and conduct of my research during the Autumn and Spring terms. They led to:

- the desire to address effectively my lack of experience in organizing, managing and teaching a Key Stage 1 class;
- the corresponding need to examine/modify my classroom practice with regard to technology with Years 1 and 2;
- the need to administer teacher- and standardized assessments;
- a need to promote effectively technology to my own school and cluster schools.

During my initial planning for my new class I was anxious to build in a significant element of group activities and tasks that would inculcate a genuine sense of cooperation between pupils, alongside of, rather than at the expense or exclusion of, individuality. This aim was primarily an ideological one, reinforced by practical necessity. I judged that in order to deliver any programme effectively and efficiently to such an unfamiliar spread of age and ability I'd need to cultivate complementary qualities of independence, initiative, self-sufficiency and cooperation from the outset of the autumn term. In these respects I was operating within the parameters outlined by the Schools Council: 'Small schools with too few pupils to form separate classes for each group may have no alternative to some vertical grouping with children of more than one age group in a class' (Schools Council 1983: 132). I envisaged that cross-curricular problem-solving activities of varying complexity geared toward design/technology would present ideal situations in which to introduce and foster the skills and attitudes necessary for successful groupwork that would, in turn, enhance the children's existing attributes. In striving to achieve this I felt I was responding to Johnsey's plea for cultivating inventiveness and initiative within children: 'Practical problem solving is the ideal setting in which children can learn to co-operate with each other' (Johnsey 1986: 10).

In the National Curriculum Programmes of Study it is expected that pupils should discuss their ideas, plans and progress with each other, and should work in groups as well as individually. This kind of collaborative work is expected to stretch into evaluating how they go about their work, and suggesting modifications to it. Within the technology Orders specific references to such activities are scattered throughout the Statements of Attainment, Programmes of Study and Non-Statutory Guidance. This is not a new idea: '[an] important characteristic of designing is that, although it can be an individual act, it has a strong, co-operative aspect. ... Designing requires discussion, collaboration and taking account of other people's needs and wants' (Design Council 1987: para 3.4 and para 4.3.).

Contrary to my original intentions the Autumn term unfolded into a series of sporadic attempts to commence my planned programme. A significant majority of the children (some of the Year 1 children were only in their second term of school, full-time) did not respond positively to what I was asking of them. I had grossly underestimated the range of ability and experience, and how I should cater for it. The children's responses illustrated their confusion about how best to do what I'd asked of them. The disparity between what the children could cope with and my teaching expectations, previous experiences, and style of delivery was considerable. I soon resolved not to begin groupwork in earnest. My priorities were adjusted to establishing an effective system of routines wherein the children, I and other adults – welfare assistant and parent helpers – could all operate confidently.

The technology content was, at this point, accordingly highly structured, teacher-directed and prescriptive, but geared toward encouraging individual responses (Bonfire Night posters, pop-up Christmas cards). The need to introduce and/or provide practice in the fundamental design/technology skills of 'making' i.e. cutting, joining, decorating, finishing, within safe working practices, became a paramount concern.

By January, whilst there were still misgivings as to the technical ability and skills mastery of the children, there were enough indicators of the overall self-sufficiency of the class to commence groupwork as originally planned.

From this point I made a concerted effort to research what happened and to match the research methods (as outlined later) to my classroom management and schemes of work. By now I was especially interested in how the children would select working partners and/or create working groups for problem-solving activities. The implications of the principles of group-work identified in the policy documents were becoming clearer to me: collaboration; cooperation; taking account of others; articulating their ideas and progress, sharing plans, and so on, are high-order expectations. I wanted to know, among other things, on what basis the children formed their working relationships and shared evaluation criteria. In particular I wondered what their ideas about designers were, and if these played a part in their choices of working partners.

In considering the methods and conduct of the enquiry it became clear to me that, in the light of the traumatic events of the Autumn term, and taking account of the additional commitments on the timetable (i.e. teacher/statutory assessments, Spring term production at half-term, Parent-Teachers Association functions, and residential trips) there would be little room for exhaustive prior evaluation of the range of research methods available. The way I understood common sense theory and knowledge in action (Bassey 1990) would tend to predominate. However the attraction to me of action research was that in this context, as an action-researcher I'd have the autonomy to set the agenda, collect data, reflect upon the data collected, formulate the course of action based upon those reflections and carry them out in the classroom.

I set the teaching agenda for design/technology activities, in ways which might help me ascertain to what degree the Year 1 and 2 children in my class possessed sufficient design-orientated capabilities to enable them to exercise an informed choice when selecting work partners. I attempted to determine how best to collect the relevant data. In essence I was concerned with two questions:

- how do I obtain data for which the ambiguities of interpretation are reduced to the lowest possible degree?
- given the purpose and question of the research, what kind of investigative exercises, operations and strategies should I embark upon to fulfil the purposes and answer the questions?

(Willens and Raush 1969: 3)

In view of the lack of formal writing skills of the majority of pupils, and in order to ensure access to a meaningful response from as many of my class as possible, much of the data was gleaned by using taped interviews (and transcripts), photography, documentary evidence, narrative records of my reflections, and observation notes of events.

What follows is a chronological résumé of the project in which I have set out the key events as they occurred during the Spring and Summer terms.

January Identification of the initial focus for research, setting-up of the trial run introductory task, and pupils photographed as they worked. The children were asked to draw a picture of a designer.

February Initial task worked through in groups and related data gathered, with additional photographs taken prior to interviews with children. Groups were of three or four children, mixed-sex, mixed ability and mixed Years 1 and 2.

This activity was the first in which the class worked in groups within a given timescale (one afternoon plus the following morning to finish off) in order to achieve a prescribed end-product, i.e. a wall chart. Its function was to show/describe the characteristics of building materials, using a variety of media for the illustrations, together with appropriate brief, written descriptions. The finished wall charts then formed part of a display to support the Spring term 'Homes' project.

February 4th Tape recording of teacher-led, whole-class discussion of children's perspectives of designers and their work, together with attributes necessary for successful designing.

Weeks commencing February 11th and 17th During this fortnight the children were asked to respond to the task of constructing a life-size model of a milk carrier, capable of carrying four or six milk bottles/cartons securely and safely from doorstep to kitchen. An indicator to show the daily order was to be an integral part of the design/model. They were asked to choose a single partner for this task.

The children selected their own partners but no reference was made to our previous discussion of positive attributes in designers. Having organized the pairs into viable working spaces, the children were asked to draw a sketch-plan of their proposed milk-carrier. Then ideas from these initial drawings were amalgamated (with an emphasis upon mutual accommodation of ideas), in order that each pair could produce a joint plan as a basis for their model. Quite independently, some children suggested labelling their sketch-plans to indicate constituent parts (and, later, the actual materials to be used in constructing the model), and these plans were combined with written/illustrated evaluations.

Following this design stage I introduced a range of materials for the children to utilize in the manufacture of a model of their milk-carrier: art-straws (jumbo/standard), A3 plain white card, paper fasteners, PVA glue and Sellotape.

March 27th The children were asked to consult with partners in evaluating their own artefact, and make a decision as to the first, second and third most successful milk carriers. This was effected by giving each pair 10 counters to be allocated as follows:

5 counters to the best artefact;
3 counters to the next best;
2 counters to the third best.

The following is a record of the total number of counters voted to each artefact:

Designers	Merit score
James/Adam	23
Louisa/Sarah-Jane/Kirsty	22
John	14
Richard/Tom	12
Stuart/Michelle	12
Matthew/Georgina	10
Kevin/David	9
Heidi	8
Lorna/Nikki	8
Thomas/Dominic	7
Daniel/Adrian	7
Sophie/Georgina	6
James/Oliver	4
Amanda/Rebecca	2

March 30th During the ensuing class discussion the criteria which emerged as to how the children had allocated their counters included references to 'well made', 'strong', 'clever' and 'nice to look at'. There was a strong consensus as to the appropriateness of such terms. Those artefacts attracting a low total were seen as being 'not good', 'not interesting', 'not finished', 'too many bits' and 'messy'. Here again there was mutual acceptance of these terms, including the makers of those models the comments were directed at.

Following this exercise the artefacts and design sheets were assembled in a display. I approached each design partnership with a view to teasing out what each child had expected of their partners, i.e. the attributes they considered most important when working with others on a design. My summary of the children's post-task comments regarding criteria for partner selection follows:

Thomas (Year 2) and Dominic (Year 2) had no choice; they worked with each other as a result of their return to school following an absence after everyone else had started.

James (Year 1) was seen by Adam as a hard worker and not bossy. Adam (Year 2) was said by James to be 'good at drawing' and 'helpful'.

John (Year 1) saw Heidi (Year 2) as helpful but also the only one left to choose. Heidi thought John was good at making things. (In the event John and Heidi worked in parallel with each other, rather than in partnership.)

Daniel (Year 2) played with Adrian (Year 1) and saw him as good at making things. Adrian also commented that Daniel was good at making things.

James (Year 2) chose Oliver (Year 1) because of his drawing ability. Oliver chose James for the same reasons.

David (Year 1) opted to work with Kevin (Year 2) because of his spelling skills. Kevin admired David's skill in joining things, i.e. his practical skills.

Matthew (Year 1) played with Georgina (Year 2) at breaktimes. Georgina worked with Matthew because he's popular.

Stuart (Year 2) opted to work with Michelle (Year 1) because she was patient. Michelle thought Stuart was good at spelling and good at making 'stuff', i.e. Lego, Clixi, Bauplay, etc.

Amanda (Year 1) linked with Rebecca (Year 2) on a friendship basis and had watched her tackling puzzles. Rebecca felt Amanda was good at colouring and making things.

Nikki (Year 1) believed Lorna (Year 2) to be helpful and not argumentative. Lorna decided that Nikki was good to work with, good fun and hard working.

Sarah-Jane felt both Louisa and Kirsty (all Year 2) would be helpful. Kirsty thought Sarah-Jane considerate, not bossy, and Louisa good at spelling and making things. Louisa felt Sara-Jane and Kirsty were both good fun and enjoyable to be with.

Sophie (Year 1) felt Georgina (Year 2) to be good at making things. Georgina said of Sophie that she was helpful.

Tom (Year 1) had watched Richard (Year 2) making things, and he was also his best friend. Richard appreciated the fact that Tom didn't argue and was good at making things.

When faced with ordering, collating and interpreting the data in writing up this project I was drawn to reflect ruefully on the following extract:

> When researchers conduct research, they attempt to answer questions in a systematic way. The answers they eventually find are as much a function of the questions asked as they are of the procedures used to

collect, analyse and interpret the evidence. Poorly formed questions will usually result in inadequate answers.

(Andersen and Burns 1989: 86)

I experienced increasing difficulty in the midst of all else that was happening to maintain an objective, critical overview of the research's progress, and plan subsequent steps. I don't wish this to translate as an excuse for the sporadic management of the research and its subsequent recording. It's merely a reflection on my attempt to conduct action research in my specific (though not, I suspect, untypical) situation. My initial response to the data and observations was to recognize my very limited experience in the intricate process of phrasing questions and answers when in discussion with Early Years children. The dangers of compounding misunderstandings, or wrongly assuming mutual understanding of the concepts under discussion, became evident. This was illustrated by the instance of the children's criteria for choosing work partners. Judged superficially in the heat of the teaching/learning moment, these appeared to be much more diverse than criteria which I might have anticipated or hoped for: clearer, more specific capabilities, and more particularly characteristics which might help teamwork, with views about how to achieve the best results possible. For a preliminary benchmark against which to assess the children's views of designers/designer capabilities I used an outline of key skills for design and technology which identified some fundamentals of 'success' in planning and making activities. In addition, capabilities were identified as being appropriate to aspects of group problem solving:

Making skills
cutting
mixing
joining
finishing

Designing skills
sketching
constructing models
drawing
communicating

Researching
classifying
evaluating
interpreting
predicting

Recording
drawing
observing
selecting information
writing

Planning
analysing
coordinating
managing
ordering

Group skills
listening
talking
negotiating
questioning

Table 11.1 Infant children's perception of designers' attributes

Interview	Designers designing pictures	Reason for partner choice: technology task evaluating
Kevin (Year 2) Reference to clothing as external indicator of 'smartness' No appropriate correlation	Good at sticking, sewing, planning colours *Making*: joining, finishing *Researching*: classifying *Recording*: observing	Admired partner's skills in joining things *Making*: joining
Adam (Year 2) No contribution recorded	Good at painting *Making*: finishing	Looked for good worker, someone 'not bossy' *Group skills*: negotiating
Stuart (Year 2) No contribution recorded	Good at drawing animals/buildings *Designing*: drawing/sketching *Recording*: drawing, observing	Wanted partner who was patient *Group skills*: listening, negotiating
Dominic (Year 2) Comments suggested designers need to be safety conscious and know about appropriate materials *Researching*: evaluating	Good at making things, testing things, writing and taking care of things *Making*: all *Researching*: evaluating *Recording*: writing *Planning*: managing	No positive opinion expressed
Daniel (Year 2) Reference to clothing as external indicator of 'smartness' No appropriate correlation	Good at painting, making things with wood *Making*: all	Partner was good at making things *Making*: all
James (Year 2) Oblique reference to planning ahead *Researching*: predicting	Good at painting/drawing *Making*: finishing	Highlighted partner's drawing ability *Recording*: drawing
Thomas (Year 2) Good at doing pictures *Designing*: sketching, drawing *Recording*: drawing	Good with colours, measuring, putting things together *Making*: joining, finishing	No positive reasons expressed
Lorna (Year 2) No contribution	Good with tools and wood *Making*: all	Partner was good to work with, good fun and hard working *Group skills*: negotiating
Kirsty (Year 2) Good at doing pictures *Designing*: sketching/drawing *Recording*: drawing	Be able to colour/write neatly *Designing*: sketching, drawing *Recording*: drawing/writing	Partners were considerate, good at spelling and making things *Group skills*: negotiating *Making*: all

Early years, designers and partner choice 127

Table 11.1 cont.

Interview	Designers designing pictures	Reason for partner choice: technology task evaluating
Heidi (Year 2) No contribution	Good at lots of things	Partner good at making things *Making*: all
Louisa (Year 2) Good at making clothes Oblique reference to outward appearance and academic ability *Making skills*: all *Planning*: managing?	Good at drawing and painting *Designing*: sketching, drawing *Recording*: drawing *Making*: finishing	Partners were good fun and enjoyable to be with *Group skills*: negotiating?
Adrian (Year 1) No contribution recorded	Designers good at designing pictures *Designing*: drawing/sketching *Recording*: drawing	Partner was good at making things *Making skills*: all
Oliver (Year 1) No relevant contribution recorded	Skilled, trained, likes art, makes things, good at everything *Making skills*: all	Chose partner on basis of drawing ability *Designing skills*: drawing *Recording skills*: drawing
James (Year 1) No contribution recorded	Good at tennis!	Partner good at drawing and helpful *Designing skills*: drawing *Recording skills*: drawing *Group skills*: negotiating
Matthew (Year 1) Good at drawing *Designing*: drawing *Recording*: drawing	Good at making things and working things out *Making*: all *Recording*: evaluating *Planning*: all	Playtime friend selected
Tom (Year 1)	Good at drawing, colouring making *Designing*: drawing *Recording*: drawing *Making*: all	Partner good at making things *Making*: all
John (Year 1) Good at making things [Knows] how to use tools *Making things*: all	Good at drawing pictures, making things, especially good at job, especially good at one thing *Designing*: drawing *Recording*: drawing *Making things*: all	Partner helpful *Group skills*: negotiating?

Table 11.1 cont.

Interview	Designers designing pictures	Reason for partner choice: technology task evaluating
David (Year 1) No relevant comment	Good at painting, using a computer, writing and making things *Making*: all *Recording*: selecting information, writing	Partner had good spelling *Recording*: writing
Amanda (Year 1) No comment recorded	Good with materials, good at making *Making*: all	Partner chosen through friendship and skill at solving puzzles *Planning*: all
Georgina (Year 1) No comments offered	Patience *Group skills*: negotiating?	Partner needed to be helpful *Group skills*: negotiating
Sophie (Year 1) No directly relevant comment recorded	Knowledgeable about materials (paraphrased) *Making*: all	Partner good at making things *Making*: all
Michelle (Year 1) No comment recorded	Good at mending/fixing *Making skills*: all	Partner good at spelling and making things *Recording*: writing *Making skills*: all
Nikki (Year 1) Good at painting *Making skills*: finishing *Recording*: observing	Good at drawing carefully, putting things in right places *Designing*: drawing *Recording*: drawing *Making*: joining	Partner helpful (i.e. not argumentative) *Group skills*: negotiating

I used this outline as a check list against which to compare children's references to designers, designing and design skills. In order to search for consistency within the range of children's comments I cross-referenced from each child (where appropriate) as follows. I paraphrased comments made from my second taped interview; alongside these I placed the relevant written comment(s) taken from the children's pictures of designers designing; and I compared these with the children's comments made during the evaluation session.

These were then tabulated (see Table 11.1). The incidence of a positive match of designer skills referred to (directly or indirectly) by the children suggest this particular class of Early Years/Infant children showed a good deal of common ground, almost a 'collective' image of the attributes and skills required by designers, and that my initial interpretations about diversity were too hasty.

However, within this 'corporate awareness' there was room for a range of possible interpretations of the evidence of perception/appreciation. Could James's (Year 1) belief that designers were tennis players indicate an increasing element of designer consciousness within the sporting world, especially televised sporting events? It is difficult to know how to interpret such ideas, which seem 'disconnected' from the question in hand. Heidi (Year 2) stated that designers needed to be good at lots of things. What should one read into such a non-specific response?

Dominic and Louisa (both Year 2) offered a broader, more sophisticated appreciation, or perhaps a more detailed articulation, than most. For the majority, 'insight' seemed erratic with regard to choice of qualities in partners. Rudimentary analysis seemed to show only tenuous evidence of a link between appreciating the qualities that make designers function effectively, and actively seeking those same qualities in work partners. In other words, for some reason they were unable or failed to register the practical opportunity for selecting work partners on the basis of design oriented skills, qualities and characteristics.

I found the data, or at least my interpretations and analyses, frustratingly untrustworthy, a feature, I suspect, of much action research. These particular infant children articulated a fairly consistent set of skills and attributes that were applicable to designers and designing activities. It does seem however that in forming groups for problem-solving tasks, there needs to be acknowledgement of group-working skills, and how those qualities and attributes can help in such work. Activities/experiences that will enhance the children's personal qualities and capabilities that are utilized when working in groups need to be self-consciously built around the development and deployment of those attributes.

In the course of conducting my research there appeared no easily identifiable strategy for developing individual contributions to group tasks. Nor do I know of any method for taking them into account in an easy-to-use, accessible manner for assessment purposes. Indeed it may be an impossible issue to resolve. Yet the practice of organizing the class into groups is common in schools and inevitable in small ones (DES 1992).

Perhaps there needs to be some method of engineering groups to contain complementary strengths and weaknesses, and design and technology activities are one forum for this to take place. I feel I've only just begun to address a fraction of the issues raised, and need to continue to re-evaluate these in the light of fresh concerns and questions arising out of those issues.

FURTHER READING

Andersen, L.W. and Burns, R.B. (1989) *Research in Classrooms – The Study of Teachers, Teaching and Instruction*, London: Pergamon Press.

Bassey, M. (1990) 'On the nature of research in education (Part 1)', *Research Intelligence*, BERA Newsletter, Summer 1990.

Department of Education and Science (DES) (1992) *Curriculum Organisation and Classroom Practice in Primary Schools: A Discussion Document*, London: HMSO.

Design Council (1987) *Design and Primary Education*, London: The Design Council.

Johnsey, R. (1986) *Problem Solving in School Science*, London: Macdonald Education.

Johnsey, R. (1990) *Design and Technology Through Problem Solving*, London: Simon & Schuster.

Kimbell, R.A., Stables, K., Wheeler, A.D., Wozmak, A.V., and Kelly, A.V. (1992) 'The assessment of performance in design and technology' in *Design and Technology Times*, Spring 1992, University of Salford Technology Education Department Unit.

Schools Council (1983) *Primary Practice – A Sequel to The Practical Curriculum: Work Paper 75*, London: Methuen.

Willens, E.P. and Raush, H.L. (1969) *Naturalistic Viewpoints in Psychological Research*, New York: Holt, Rhinehart & Winston.

Chapter 12

Technology teaching at Dove First School

Gillian Oliver

Prior to my move within the school to teaching a Year 2/3 class, a technology specialist teacher taught the subject to all the Year 2 and 3 pupils. Faced with the loss of this resource we embarked on a journey into the unknown which involved some staff teaching technology for the first time, and myself in coordinating a new subject. The change in staffing and the fact that some members of staff were not happy with the swapping of classes involved with specialist teaching led us to take the decision to teach technology with our 'own' classes. Specialist teaching also had the disadvantage of restricting technology teaching to a fixed time slot, rather than integrating with the topic or allowing flexibility if a project required extension.

The school has two reception classes, two Year 1 classes (one with six younger Year 2s) and three Year 2/3 combined classes. We have mixed ability classes and employ a range of teaching styles and methods to suit particular curriculum activities and the needs of our children. Our curriculum includes a whole-school, four-year rolling programme of term-long topics, which we consider to be flexible enough to allow us to incorporate all areas of the curriculum. (We are aware that some topics lend themselves more easily to a scientific or a humanities bias.) We assess the children in the form of an 'I can do' record of achievement, which in the case of technology was prepared using the national curriculum orders.

Impressions of what was being produced during the previous year suggested that the children were experiencing a limited range of activities. There seemed to be a considerable difference between what I had described in the policy document and what was happening in practice. In particular I was concerned about 'progression' from one year group to the next. The National Curriculum Council considered the importance of progression in the (1993) Technology Order review: 'The Council has analysed progression between statements of attainment at different levels and has amended statements to clarify progression and improve accessibility' (NCC 1993: 7). What I perceived as a lack of technological development in school could have resulted from a variety of causes, which I postulated and decided to investigate. My research aimed to illuminate:

Resources

- Were the resources in school adequate for the teaching of technology for children in the 4–8 year age range?
- Were staff aware of the resources available and their location in school?
- Were the staff happy using the resources available with the children?
- Did they feel that they had sufficient knowledge available for their use?

Curriculum knowledge and skills

Were the staff confident about which knowledge and skills should be taught to the age-range for which they were responsible?

Other constraints

- By allowing staff the freedom to comment on any other constraints that affected their teaching, could I discern other issues on which I might need to act as coordinator, or which might explain what I perceived to be the situation?

I set out to investigate these areas in our school with the cooperation of the staff. Initially though, I set out to investigate the technology being taught, in order to gather evidence which might support or refute my impressions about the technological experiences of the children. I set about the research using:

Staff interview aimed to develop an understanding of the way staff were approaching their technology teaching. It was semi-structured to gain insight into planning, their approach to individual activities, and resourcing.

Questionnaire and analysis grids using the issues raised at interview I investigated aspects of resourcing and the skills being taught at each age range.

Teacher confidence ratings to investigate the level of confidence teachers had in delivering various aspects of technology.

The data collection and analysis was intended to help me formulate actions required by my role of coordinator.

Dove School serves a small market town and five nearby villages. All of the classrooms have enough space to allow flexible organization, to suit

Table 12.1 Materials grid

Year group	Materials	Request
R (2 classes)	Scissors Glue Paper Big Builder Art-straws Constructo-straws Duplo Mobilo Sticklebricks Blocks Octogons	Glue gun Fabric paints Pearlized paint Fluorescent paint Straws
1 & 2 (2 classes)	Scissors Glue Paper Polydrons Blocks Sticklebricks	Lego
2 & 3 (3 classes)	Paper Card Glue Lego Mini Quadro Unifix Wool Felt Binca Weaving cards Wood Saws Sawing blocks Drill Art-straws Dowel	Glue gun Multilink cubes Lego technic Mini vice More saws Fabric crayons Fabric paints Roller and tray Silk screen Printing blocks

the needs of the whole primary school curriculum. Seven full time teaching staff (at varying stages of their teaching careers) are helped by one full time welfare assistant and three part time welfare staff.

I had inherited the classroom of the specialist teacher, which was well resourced with a range of equipment and materials, but I was aware that either these required distributing to other classes, or their equivalent needed to be made available in order to allow other classes to optimize technology teaching. To this end I set about investigating what resources staff had in their classrooms, and what they would request given the opportunity to invest further. I collected data from each class teacher using a simple resources questionnaire. The findings from these investigations revealed that throughout the school our technology resources are very limited (see Table 12.1).

Conducting interviews with colleagues was a very sensitive matter, especially when they were at the early stages of technology teaching. I made notes at the end of the interviews, which provided three key points:

- staff were aware that they had not included some basics in their resource lists;
- there was confusion over what was technology/art/science;
- staff were unaware of what was available in other classrooms.

I had invited suggestions for the acquisition of resources, and was surprised by the lack of requests for further equipment (Table 12.1). Exploration of this at interview suggested two possible reasons:

- staff were aware of the current financial situation and had made their requests accordingly;
- staff were unaware of what resources were available on the market and available for use with their pupils' age group.

As a result of these interviews I decided to investigate the use of resources in more detail, breaking down technology materials into recycled materials; construction materials; construction kits; textiles; food.

I also wanted to look at the knowledge and skills that each teacher was teaching to their pupils, to develop my understanding of how these resources were being used in each age range. If new materials were being introduced as the children progressed from Year R/1 to Year 2/3 then this provided some evidence for technology 'progression' in that the children were experiencing working with an increasingly broader range of materials.

Analysis of Tables 12.2 to 12.6 shows that few new materials are introduced to the children at each age range and that some materials cease to be available as the children move from Year R to Year 1 and from Year 1 to Year 2/3. More materials and construction kits are withdrawn from use than are introduced (see Table 12.6). The children experienced a very big reduction in the number and type of construction kits that were available to them as they moved from Year R to Year 1 and then to Year 2/3 classes. Seven construction kits previously used were no longer available to the children when they entered a Year 2/3 class.

Evidence of the use of food proved rather difficult to analyse. No member of staff had a regular preference for cookery activity. Children do not therefore experience working with standard materials and recipes that are common to their year group, and known to other teachers. Teachers particularly in Year 2/3 reported that they chose their food work to fit in with their topic, allowing the children to experience a range of foods. This is an area where the knowledge and skills being taught will

Table 12.2 Year R materials

Recycled materials	Construction materials	Construction kits	Textiles	Food
Junk boxes	Paper	Duplo	Cottons	Flour and salt dough
String	Cardboard	Lego	Polyesters	Chocolate
Wool	Wallpaper paste	Mobilo	Fur fabric	Flour
Paper plates	Marvin	Start Gear	Binca	Fat
Apple trays	Plasticine	Bricks	Anchor Perle	Eggs
Toilet rolls		Big Builder	Blind Samples	Cornflakes
Cylinders		Constructo-straws		Crispies
Wallpaper		Sticklebricks		
Cotton sheets		Popoids		
Off-cuts of materials				

Table 12.3 Year 1 materials

Recycled materials	Construction materials	Construction kits	Textiles	Food
Toilet rolls	Toilet rolls	Lego	Wool	Flour
Straws	Straws	Start Gear	Fabrics	Sugar
Boxes	Boxes	Playmobile	String	Fat
Cylinders	Cylinders	Constructo-straws	Felt	Cereals
Card	Card	Polydrons	Binca	
Paper	Paper	Lego Dacta	Threads	
Newspaper	Newspaper	Meccano		
Plastic bottles	Plastic bottles	Big Builder		
	Plasticine			

Table 12.4 Year 2/3 materials

Recycled materials	Construction materials	Construction kits	Textiles	Food
Boxes Plastic bottles Plastic bags Newspaper Card String Toilet roll middles Kitchen roll middles	Glue and adhesives Split pins Paper clips Tape *Wood *Card wheels *Corraflute *Syringes *Plastic tube *Balloons	Lego Technic Mobilo Polydrons Unifix cubes	*Felt Dyes Cotton Wool *Fabric crayons *Fabric paints Binca	Ingredients for topic based cooking

Key * Class 4 based resources

Table 12.5 New materials introduced at change of age range

New age range	Recycled materials	Construction materials	Construction kits	Textiles	Food
Year 1	Plastic bottles	—	Playmobile Lego Dacta Meccano Lego Technic	—	—
Year 2/3	Plastic bags	Split pins Paper clips Corraflute Card wheels Wood		Dyes Fabric crayons Fabric paints	As per topic

Table 12.6 Materials no longer available at change of age range

New age range	Recycled materials	Construction materials	Construction kits	Textiles
Year 1	Paper plates Apple trays Wallpaper Cotton sheets Off-cuts of material	Wallpaper paste	Bricks Sticklebricks Popoids	Cottons Polyesters Fur fabric Blind samples
Year 2/3	–	–	Start Gear Playmobile Constructo-straws Lego Dacta Meccano Big Builder	–

Table 12.7 Year R skills

Recycled materials	Construction materials	Construction kits	Textiles	Food
Cutting Sticking Painting Fixing Talking Investigating properties Drawing objects they have made	Cutting Sticking Painting Fixing Talking Investigating properties Drawing objects they have made	Following patterns Copying a model Designing Adjusting/modifying Talking Evaluating	Sensory skills Selection Description	Mixing Stirring Beating Rolling Cutting Hygiene Texture Following instructions

Table 12.8 Year 1 skills

Recycled materials	Construction materials	Construction kits	Textiles	Food
Cutting Sticking Problem solving Drawing Designing	Cutting Sticking Problem solving Drawing Designing Moulding	Coordination Problem solving Using gears Pushes and pulls	Using tools, e.g. needles Sewing Cutting Sticking Appliqué	Following instructions Modelling Texture

Table 12.9 Year 2/3 skills

Recycled material	Construction materials	Construction kits	Textiles	Food
Drawing Sticking/fixing Measuring Hammering Cutting Screwing Sorting Sawing Using glue gun	Drawing Sticking/fixing Measuring Hammering Cutting Screwing Sorting Sawing Using glue gun	Following instructions Drawing Fixing Learning about systems and mechanisms	Sewing Weaving Textile printing Painting Dyeing	Weighing Mixing Heating Hygiene

Table 12.10 New skills at each age range

New age range	Recycled materials	Construction materials	Construction kits	Textiles	Food
Year 1	Problem solving Designing	Problem solving Designing Moulding	Problem solving Using gears Pushes and pulls	Using tools Cutting Sticking Appliqué	Modelling
Year 2/3	Hammering Screwing Sorting Sawing Using glue gun	Hammering Screwing Sorting Sawing Using glue gun	Systems	Weaving Printing Painting	Hygiene Heating

Table 12.11 Skills teaching ceasing at each age range

New age range	Recycled materials	Construction materials	Construction kits	Textiles	Food
Year 1	Painting Talking Investigating properties	Painting Talking Investigating properties	Following instructions Copying a model Talking Evaluating Modifying	Sensory skills Selection Description	Mixing Stirring Beating Hygiene Rolling Cutting
Year 2/3	Problem solving Designing	Problem solving Designing	Problem solving Coordination	Cutting Sticking Appliqué	Following instructions Modelling Texture

need to be looked at more closely, as the same skill could be taught with a wide variety of ingredients. One teacher did not report any food work at all.

The use of construction materials, particularly in the early years, is very limited. This could be due to substitution with recycled materials and construction kits. We certainly need to consider the range of materials available to the children, and whether substitutions, where they occur, are appropriate. If kits are used in place of construction materials children could be denied the opportunity to develop some basic skills. Perhaps equally importantly, they may be denied the 'ownership' of what they have produced.

I was concerned about the uneven distribution of the use of materials across the age ranges. I postulated that this could be due to the historical factor of Class 4 previously being used as the technology room for Year 2 and Year 3 pupils. I needed to make staff more aware of what is available both within school and available in the catalogues, to ensure full use of our limited resources and to extend them.

Evidence of access to materials is not necessarily sufficient evidence for the teaching of advanced technology skills, however. What I needed to investigate next to allow me to address the question of what range of technology teaching was happening at Dove was what knowledge and skills are taught to each age range.

Analysis of Tables 12.7 to 12.10 shows that the skills involved using recycled materials and construction materials are very similar for all three age ranges, suggesting that we needed to consider whether we have identified potential progression in skills. We also needed to consider whether this use of recycled materials is 'cost-led', and convenient rather than educative. Construction materials are expensive and are expended quickly when used with a class of children, but the opportunity to produce work with a high quality finish depends on access to such resources.

Table 12.10 shows the basic skills said to be introduced to each age group. As evidence of progression that is occurring this appears to be very basic, and told us little of what specific skills were being taught or to what standards. 'Problem solving' as a skill can be approached in a variety of ways, and the question of whether problem solving, for example, meant the same thing to all of us was not answered by this research. The table shows some very specific skills such as using gears and the technique of appliqué. These came from one member of staff who had planned her programme for the year using this year's topics. These are skills which she, independently, thought appropriate to those particular topics. They might not be extended.

It does, however, provide a starting point for future discussion and planning/development, for designing curriculum progression, detailing the basic skills that should be covered by each year group so that our technology teaching builds in a coherent way on the experiences of the children year-by-year. The need to do this became especially evident from Table 12.11, which shows the skills and activities ceasing to be identified by teachers at each age range beyond Reception. The fact that some skills and activities were no longer identified by teachers could be due to a number of factors:

- It was assumed the work had been grasped and was no longer targeted. Did this mean that the skills were no longer practised or practised but not explicitly identified?
- The skills were practised elsewhere, e.g. following instructions can appear in various aspects of technology as well as other curriculum areas. Were skills explicit to technology or could they appear in related curriculum areas such as science and art?
- Skills can be identified in different forms. For example cutting may not explicitly appear as a Year 2/3 skill but sawing, a particular kind of cutting skill, may be identified.

This meant that it was necessary to consider the teaching of skills and how we practise those skills, in ways which allow for pupils who have not grasped a skill in the specified age range to grasp and rehearse that skill in the next age range. If the third postulation applies we may need to consider the links with other subjects to see if we are covering these skills under a different curriculum heading.

TEACHER CONFIDENCE

I considered the confidence of the teachers involved in teaching technology, using five aspects of 'content' from the Programmes of Study. I asked staff to rank their 'teaching strengths' on a seven point scale from 'need to develop confidence' to 'confident'. Only four of the seven teachers completed the questionnaire (Table 12.12). The construction of the table was helpful in developing a list of staff training and development needs. The lack of staff confidence is very important and these indicators

Table 12.12 Teacher confidence analysis scale

		Materials	Mechanisms	Structures	Textiles	Food
Very confident	7	–	–	–	1	1
	6	3	–	–	–	1
	5	–	–	–	–	–
	4	1	1	3	2	1
	3	–	1	1	–	–
	2	–	2	–	1	1
	1	–	–	–	–	–
No confidence	0					

suggested that we would be wise to consider the area of staff technology training on the School Management Plan to be addressed in the future.

When questioned on the factors limiting the scope of their technology teaching staff identified various restrictions:

- 'time and the degree to which the children need to be supervised during some technology activities';
- 'time and the chance to talk through group work with the children';
- 'time and numbers (of children), I have voluntary help but I use it for reading skills';
- 'lack of materials in school and access to those materials'.

Staff also indicated a desire for project packs where the planning and material is already available. We have begun to develop a resource bank of 'ideas', but from talking to staff I felt that they were hoping for kits equivalent to the science and new Lego Technic kits we have in school. The science and Lego Technic kits provide activities to extend and develop pupil learning through permanent and re-usable components. The problem with creating our own kits that I felt they had not appreciated was the question of the 'consumable' nature of technology materials. Where pupils need to design and make their own items and explore the materials available, and to experience ownership of what they have produced, any kit might be quickly exhausted.

The need for projects and activities that the children could undertake independently was also highlighted. Staff indicated that they would like technology INSET but acknowledged that their class priorities were with language and number skills, so that pupil activities which did not demand teacher attention would be desirable. The need to supervise activities or to have a 'helper' present when technology activities are taking place was also a major concern for staff. Here it may be useful to look at how we are currently managing technology in the classroom. I know from discussion that the teachers in school employ a variety of teaching methods from group to whole class approaches in various areas of the curriculum. Sharing our existing approaches to 'high contact' activities may be helpful. It may also be possible to attract more parental help into school. This was closely related to the question of the time available in the curriculum, which featured heavily among the list of restrictions identified by staff. This is a difficult area to address but I am hopeful that the modified national curriculum will lessen the obvious concerns expressed by members of staff. It may also be helpful to highlight the areas of overlap between subjects, in particular those of science and art where teachers are also covering technology-related skills.

JOURNEY OF CHANGE

All three of the factors I identified at the beginning of the study as limiting our technology teaching are influential. The question then arises as to the best way to begin our journey of change. For me the obvious starting point is to begin to address the question of resources, as identifying activities and skills to be taught without addressing the resources available would, in my view, only increase staff frustration. Once we have begun to address the resources in school we can use the skills analysis as a basis for deciding what skills should be introduced to each year group. This will help us provide a framework so that children in the same year group but different classes receive the same curriculum experiences.

The area of teacher confidence showed that INSET provision is vital in raising teacher knowledge. However not all members of staff would need or want to attend workshops on all aspects of subject knowledge, so these would be best delivered on a voluntary basis after school. Discussion with staff will be necessary. Year 3 after-school workshops on different aspects of technology could provide a 'joint' extra-curriculum activity for children and in-service opportunity for teachers.

Having undertaken the review, an action plan which included colleagues in decision making was devised, in order to achieve 'collective change' within which each teacher would have a whole-school view of developments. This was begun by raising a series of questions, intended to lead to decisions and actions, and further review on each of the constraining factors. It began:

As a whole staff we need to consider and address the following questions:

- **Resources** We now know what we have in school. Where are we going to keep our resources? How are we going to store them? Do we need a booking system? What resources do we need to teach the areas of technology identified across the age ranges? What is available in catalogues to help us? What do we need to purchase? What are our purchasing priorities?
- **Developing knowledge and skills** How are we going to develop our knowledge and skills in the five areas of technology identified? Are we going to devote INSET days, staff meeting time or after school workshops to technology skills development?
- **Using help** How can we extend what we want to achieve with the help of welfare staff and parents?
- **Review** What will be a suitable review timetable for investigating what we are teaching to see if we have achieved our aim of developing technology throughout the school?

It was important that this basis for an action plan was presented to the staff in the form of questions, to arrive at the answers together, to establish mutual 'ownership' of what we are trying to achieve, and to work on achieving it together.

FURTHER READING

Bindon, A. and Cole, P. (1991) *Teaching Design and Technology in the Primary Classroom*, London: Blackie.

Dearing, R. (1993) *The National Curriculum and its Assessment*, London: School Curriculum and Assessment Authority.

Department for Education DfE (1992) *Technology for Ages 5 to 16; Proposals of the Secretary of State for Education and the Secretary of State for Wales*, York: NCC.

Department for Education DfE (1995) *Design and Technology in the National Curriculum*, London: HMSO.

Doherty, P. (1994) 'Planning for capability and progression for design and technology in the national curriculum', in F. Banks (ed.) *Teaching Technology*, London: Routledge.

Hertfordshire County Council (1993) *Suggested Progression Within Aspects of Design and Technology for Key Stages 1 & 2*, Wheathampstead: Hertfordshire LEA.

NCC (1993) *Technology Programmes of Study and Attainment Targets: Recommendations of the National Curriculum Council*, London: School Curriculum and Assessment Authority.

SCAA (1995) *Key Stages 1 and 2 Design and Technology*, London: School Curriculum and Assessment Authority.

Williams, P. and Jinks, D. (1985) *Design and Technology 5–12*, London: Falmer Press.

Chapter 13

Children's choices

Candy Rogers

This research will show how the issues of choice, decision making, and participation have been addressed through an investigation of the work of the Reception and Year 1 children I teach. It will show how the teaching strategies I employed changed as I came to realize the effects on the quality of educational experience of the children, especially of my judgements in relation to their opportunities to choose and to take responsibility.

Typically in the past I would select a topic and through discussion led by me draw the children's attention to a limited range of materials chosen by me. I would talk through the making process, allowing only fairly small margins for what I would consider failure. The results were to my eye good, even spectacular on occasions! The children were pleased with their work, and, bolstered by compliments from colleagues, I continued in this way. Displays of work were all very neat, and it was always evident in the classroom how important it was to me that a satisfactory 'outcome' was achieved. This concern became an important part of my planning process.

My class has 23 children, 6 of whom are Year 1 and 17 Reception. In order to be clear about what the children were actually doing, I decided to observe a group of Reception children as they selected materials for tasks. I watched and listened for the duration of the activities, making notes about what they did and said. Over the period of the project I came to realize one of the most interesting problems was how to offer the children a truly unconstrained choice. Whether that was actually achieved would be the topic of further observations. In the event the tasks set at a later stage were very different from those I had set out to observe. My influences on the children's choices were progressively less, and the focus of the observations became more refined. Those changes in my own actions became the focus of attention in the research as much as the way the children made choices, as I became more self-conscious about the direct relationship between the two.

There were three activities set for the children in this project, and their responses to them were observed. My diary recorded the following notes:

Task 1

Design and make a collage picture for a calendar. The children were a group of full time Reception. They were all confident and articulate, having very clear ideas. We discussed possible pictures for the calendar and looked at all the materials available. The children did not make any preliminary sketches but I asked them what they had decided to make before they began. I then observed their choices of materials and we talked as they worked. When the collage was complete I asked the children a set of questions and recorded their responses. Materials available were paper (crepe, tissue, sticky, sugar, white, thick and thin card, fluorescent), wool, assorted textiles, straws, lolly sticks, silver and gold foil, scissors, glue, Sellotape, pencils, crayons, feltpens.

Task 2

Design and make a Christmas card on the theme of 'trees'. The same group of children were involved, but for this observation I concentrated on four of the youngest children in the class (aged around four and a half years). As with Task 1 we discussed possible designs and looked at some commercial Christmas cards. During the making process I talked to the children, though tried not to influence their choices. When the cards were finished I asked questions of the four children while we looked at the card. The materials available were the same as for task one, plus embossed papers of various colours, and sequins.

Task 3

This was not so much a task as a project, initiated by a group of mixed Reception/Year 1 following the introduction of books from the library service project collection. Our topic for the half term was 'toys' and included in the collection were several books about puppet making. They were interested in several designs which we later made, but particularly wanted to try to make a puppet theatre, 'and go round all the other classes telling a story.' By the time they came to me, their ideas were already quite advanced. They had begun to sketch characters from Cinderella, and had found a cardboard box. I encouraged them to talk about their ideas and to look for materials they wanted. There was no intentional restriction on choice and some of the group went to other parts of the school to find exactly what they wanted. I watched their progress without any interference and, on completion of the task, tape-recorded an interview with each of the Reception children involved.

Children's choices 147

A class teacher's expectations of the children's work is, I believe, a force for the raising of standards. The way in which I have always worked has been based on the assumption that high expectations nearly always results in quality work. My perception of 'quality' is an important part of each project. In considering what I expect them to produce prior to the start of each task, those perceptions usually become evident to me. They tend to be reinforced as I make it clear what I expect the children to make.

TASK 1

My thoughts were that the children would not be able to stick to the subject initially decided upon and planned. Also, faced with a greater range of materials, that they would cut and stick randomly. I thought they would choose the fluorescent and metallic papers first and that there would be a certain amount of copying. It is clear that I was not anticipating a thoughtful approach towards the greater choice of materials, and indeed rather expected confirmation that the tried and tested methods used previously would prove to be the best. In fact my initial thoughts were probably rather short sighted as the children had all experienced working with these materials before in a variety of ways. The sessions were normally with an adult and focused on the skills needed with, for instance, scissors and glue. So there was a history of work not really taken into account in my 'predictive thinking'.

The specific items which they planned before the making activity began were:

KAYLEY – a rabbit made out of tissue.
WHITNEY – my baby sister made out of lots of different papers.
LUKE – a flower made out of sticky shapes.
KERRY – some flowers (not decided about materials).
SHEENA – a fish made out of wool.
HANNAH – a house made out of wool and paper.
ROSIE – a house made out of paper.

The children involved were interviewed when the picture was finished and the same questions were asked of each. In the event Kayley made a picture of a house and a tree. It was made with tissue paper, sticky paper and textiles. Kayley was uncertain how to begin and watched the other children as they looked through the materials. She spent several minutes looking at and feeling the different papers, turning them over and talking about the colours. She chose several papers and some textiles, taking them back to her place. During the activity she went back to look at the paper again and chose another batch including the tissue paper. She watched the other children and her picture changed as she worked from her original idea to that of the house and tree.

CR   Why did you choose tissue paper for the house?
KAYLEY   Because I thought it would be house coloured.
CR   Why did you choose the sticky paper for the windows and tree trunk?
KAYLEY   Because it is my favourite colour and I like licking it.
CR   Are you pleased with your picture?
KAYLEY   Yes.
CR   Would you change anything about it if you could?
KAYLEY   Yes I'd like to use a different coloured paper for the bottom bit. [Points to the base card.]

Whitney's picture was, as planned, a collage of her baby sister. She used textiles, sticky paper and felt pens. Whitney was very confident about choosing and spoke constantly during the process. She told me that she wanted pretty material for her sister's dress, and very quickly looked through the tissue paper for the face colour she wanted. Unlike the others, Whitney at no time seemed to browse but had very definite ideas and worked very quickly.

CR   Why did you choose this material [a silk-like textile] for the clothes?
WHITNEY   Because it was pretty and I wanted to make a pretty dress for Willow. [Sister's name]
CR   Why did you choose the tissue paper and wool for her face?
WHITNEY   Because the tissue looks like skin and the wool looks the same colour as Willow's hair.
CR   Are you pleased with your picture?
WHITNEY   Yes.
CR   Would you change anything if you could?
WHITNEY   No, I think it is really good.

Luke made a flower as planned. He was quite tentative at first and looked for a lead from the others. He spent about five minutes looking through the various trays and talking about the items. He often asked if he was allowed to use this or that and built up quite a collection at his table, guarding it fiercely. During the task he went through his collection of items several times, feeling them and choosing the next piece as the picture grew. In the event it was made out of sticky paper, textiles and wool 'because I liked the stickiness. I like having sticky hands.' The material and wool for the leaves and grass were chosen 'because they were the best'. Luke was pleased with his picture, although if he could change it he'd use the plastic material: 'I like plastic.'

Kerry made a flower as planned and used the sticky shapes, textiles, and tissue paper, which she chose 'because the shape made me think of the middle of a flower'. The sticky shapes and material for the petals were chosen because she wanted the petals to be 'sticky out'. When asked about possible changes she said: 'I think I would have put some decoration on it.' Kerry talked in great detail about her picture and told me that she was going to use the sticky paper. She spent the first few minutes licking and sticking the pieces together, but not on the card. After finding that this left an unpleasant taste in her mouth she asked me if I would lick it for her. There followed a discussion about alternative methods of sticking the paper but she was determined that as it was sticky paper she was going to lick it (or have someone lick it for her). When Kerry eventually started on the picture it progressed quite quickly and she seemed to have definite ideas about the materials she wanted.

Sheena was not happy about the finished picture when interviewed afterwards, but up until that point she had, in fact been very involved and determined in her attitude to the task. During the time she had, she used a greater variety of materials than anyone else and seemed happy in the exploration of the materials. However, when everyone had finished, the children talked about what they had done and showed their pictures to the class. Another child remarked, 'Look at Sheena's, what a mess.' I could almost see the confidence fall away. Sheena's picture showed no form which I could identify as representational, but she said it was a fish. She used sticky paper, wool and textiles, but 'I don't know' was the response to the question 'why?'. She was pleased with her picture, but if she could change anything she would 'make a hat'.

Hannah changed her mind and made a picture of her mum. She made a couple of false starts on her picture and pulled off the paper and textiles to begin again. Everything got rather gluey but she kept on task. She did not look at all the materials on offer though I presumed she was aware what was there as we had previously discussed them. She seemed to choose from the trays close to her table and, when asked, said that she didn't want any of the other things. She used sticky paper for the body 'because I like scrunching it up'. She choose the wool for the hair because it 'looks like hair, but I couldn't fit it all in'. If she could change anything she would 'take it all off and do a house instead'.

Rosie made a picture of a house and a tree. It was made out of textiles and tissue paper, and finished with crayons. She chose tissue paper for the house 'because it is pretty', and fabric for the leaves and tree trunk because it was 'a pretty colour'. She liked it so much she would not change anything. Rosie had very definite ideas about what she wanted and how the picture was going to look when it was finished. She spent a long time looking for the materials she used for the tree trunk and leaves. She talked about the size and colour of the leaves but was not able to cut them out with the scissors and asked for assistance. Throughout the time I didn't see her trying different materials. Her work was completed very quickly.

TASK 2

I felt this would be a good test of the children's use of opportunities to make appropriate choices. There was a range of materials to choose from, some of which I did not consider suitable for the Christmas card design. I expected some of the children to take a great deal of care and thought before deciding which to use, and recorded in my notes that 'I would be surprised if some cards did not turn out to be a "hotch-potch" of sticking'.

In some ways this was a more risky venture for me. Although the design was more planned and the focus more precise than in Task 1, there was the possibility for a greater deviation from that which I considered acceptable. I realized at this stage that the responses from the children when the work was complete would be vital in determining whether the work was indeed acceptable. There were four children working in a group without help from an adult. After discussing the initial ideas, the children worked on their own. I observed them throughout the time and, when the cards were finished, interviewed the children individually.

Gemma made her choices and was very careful about the decorations. She went through the box of sequins with great care and only chose the ones that were the same size. This was a very fiddly operation and I was surprised that she persisted. Gemma had chosen the red card and cut a tree out of the pink embossed paper. She had also used the sequins for decoration 'because it was pretty and right for a card', and the embossed paper 'because I liked it'. She 'would like to put some presents under the tree'.

Christopher chose orange coloured card and made a tree out of pink embossed foil. He decorated it with sequins and cotton wool. He chose the orange card 'because I liked the colour', and the foil because 'I liked the pattern on the paper'. Christopher talked about the pattern on the embossed foil and how it looked like his Christmas tree lights at home. He said that he wanted that colour foil because the other colours did not have the same patterns on them.

Carl did not seem to settle to this activity and was reluctant to use the glue, preferring to use the Sellotape. He appeared to be concerned about cleanliness and tidyness. This would surely limit choices. He chose the materials and seemed pleased with the result but there appeared to be a certain amount of relief when the task was over. Carl chose the red card and made two trees out of green embossed foil. He wanted the card because 'it was my favourite', and made the trees out of foil 'because I wanted to'. If he was to amend the design he 'wouldn't put the big sequin on that tree'.

Ulanda chose the blue card and made a tree from the silver paper because it was 'nice' and 'looks the best'. She seemed extremely thoughtful about her choices even though it did not take her long to make her mind up. She looked carefully at each paper and made her selections. Her finished card was very neat and she was happy with the result but I got the impression that she would have liked to try other materials as well.

This task had limitations in terms of choice and content and the children were shown all the alternatives. All made their choices very quickly and did not appear to be unduly influenced by others.

TASK 3

I really didn't believe this was an appropriate project for children of this age. Having said that, the children were so keen, and had already chosen some materials, that I encouraged them to try. I expected the children to make sensible choices of materials for the theatre and backdrops, but I was not so certain that they would be able to choose the rigid card or wood for hanging the backdrops and holding puppets. I thought some children would probably select paper for the puppets.

For me this task would normally be unthinkable with children of this age, and it was only due to their enthusiasm and the fact that it made such a wonderful topic for observation that it went on to completion with the minimum of adult interference. It had become increasingly apparent that this group were unusually able for their age, not only in design and technology but in all the other curriculum areas also. The fact that they could discuss amongst themselves when making choices, and accept the majority viewpoint (most of the time) seemed quite remarkable. However there were still doubts about the outcome and it was not with any great confidence on my part that they embarked on this venture.

The project was entirely the children's own idea and much of the work completed without adult assistance. There came a time towards the end of the 'making' when the group needed help in organizing themselves into a team for the performance, but apart from this my role was one of observer and interested party. I asked questions and talked to the children as they were working and on completion of the project, and interviewed each child about their choices of materials.

Helen made 'the ugly sisters and the wedding cake, and Cinderella and the Prince on their own, and another one of the Prince on his own, and I done the church scene'. She made the characters out of card and sticks, and used crayons and felt tips for details. The card was chosen 'because it was stronger than the other paper and things', Sellotape 'because the glue would come to pieces and break'.

Gemma made the mother and the house scene out of crayons, board and a stick, fixed together with Sellotape. She chose the board because it was 'a bit lighter than the other things' and because she liked the colour.

Figure 13.1 Unexpected result – the theatre is made

Anna made the 'inside of the house' scene, Cinderella in her best clothes dancing, and she painted the box and put up the curtains. She used card and sticks mainly, fixed by Sellotape. The card was chosen because 'it was more stable than the paper and it wouldn't rip as easily'.

Kayley made the two sisters in ball gowns, from cardboard and a stick, also fixed with Sellotape. She didn't know why she had made these choices, but was pleased with her work.

Lee made the carriage and horses, the coachman and the dog, using card, paper and sticks, Sellotape and glue, 'because the paper would rip, the stick was strong, and the adhesives 'would hold it on stiffer'. He especially liked doing the show to Class 2.

Kerry made the ballroom scene, the cat and the wedding cake, from mostly card, some paper and sticks, and Sellotape. The colour of the card 'was good and the crayons showed up well'. Sellotape was used 'because I would have to have waited for the glue to dry and I wanted to use the bits straight away'.

I have already described how the task came about and it took about three weeks to complete. The children's enthusiasm did not diminish. If anything they were more organized and keen at the end. They appeared to relish having control of the project. Anna knew from the start exactly what she wanted to do and started making the background pictures. She chose the card and felt pens for this, with a strip of balsa wood for the top. Kerry wanted to make the characters along with Helen and Gemma. We discussed which characters would be needed for this Cinderella story and Lee suggested looking in the library for a book. A more rigid card was chosen for the characters and balsa sticks for the rods. There was some discussion about where to fix the rods and Lee realized that the theatre would have to be made in order to be sure. A box had already been chosen, and Lee started to talk about where to cut holes. He referred to the puppet-making book that had originally inspired them, and drew on the top and sides of the box where I was to cut. The children did not have any means of cutting the box so this was the one time I gave practical help.

When all the characters and backdrops had been made and tested in the theatre the children decided to make some curtains for the front. This was a time when I felt certain they would not be able to complete the task. They chose a piece of fabric and worked out the size needed, but had difficulty fixing it to the box. Eventually Anna said that her curtains at home were tied up with ribbon. Helen and Gemma said they had curtains too that were tied. After that the job was finished quite quickly, glueing the curtains in place and fixing with a ribbon. The theatre turned out so well that the children were able to use it to give a performance to the rest of the class. For me it was certainly a most unexpected result, and the task with the least amount of input from me.

WHY DID YOU CHOOSE THAT?

Working with this group of Reception children has made me question my assumptions about their capabilities. I strongly felt that the correct way to proceed was to only offer limited choices, as the children would spoil work by making inappropriate selections from a wide range of materials. I also believed that it was vital to ensure a successful outcome for each item. This would build confidence and self esteem for further work, I thought. I was, in essence, exerting a strong hold over the activities and placing my values and judgements on the children's work, to the extent that there was very little room for individual creativity and certainly very little room for mistakes.

Similar control exerted over a class of eight year olds is reported in Tickle 1987. As in my case the teacher [John] also believed that he was offering the children opportunities for choice and individual judgement even though: 'the size, shape and range of materials was defined by John'. Following a subsequent conversation with John about the children's work, Tickle found:

> that direct instruction in what was required by way of subject matter and working procedure was supplemented by strategic decision-making about resources. Particular colours and techniques were predetermined by John, deliberately to ensure particular kinds of products.
>
> (Tickle 1987: 64)

Having said that, much of the work going on in my room was of great value in building up skills and experience. Perhaps it is partly because of this that the children were so confident in making their own choices during this 'test' period.

At the beginning the focus was at times blurred when considering why children make particular choices as the question of 'how' kept cropping up. It is interesting to note differences between the children's responses when involved in the task and those after completion. So my early records of how the choice was made can act as an aid to understanding why the choice was made when considering the children's responses. On the face of it the children were making the choices and in ways which appeared to be appropriate to the task. I summarized the significance of those choices from the data, under the following categories:

Colour preferences
Tactile preferences
Shape preferences
Technical properties.

Throughout the period of the research I had been conscious of the shift in my position from decision maker on behalf of the children, to facilitator for them. At the beginning I felt I was undertaking a project that would confirm the correctness of my provision and management of materials towards 'satisfactory products' for the children. By testing their reasons for choice it would become apparent, I believed, that they could not make appropriate choices for the tasks in hand. I was seeking to confirm my belief that a systematic programme of 'controlled' tasks would be best to build up the knowledge and skills needed to be able to cope with a wide range of choices. This would not be attainable until probably somewhere near the end of Key Stage One, I thought. In the event the way in which I was able to deal with the 'risk' of the children's 'failure' challenged my whole philosophy about 'success' and ownership of success. The children were confident and mostly articulate about their choices. Very few wanted to make changes to what they had done, and therefore inferred that the work was successful in their judgement.

The children do need to build up skills, have a variety of experiences using various materials, and opportunities to test their ideas. I believe now that the most effective way to do this in the time available is to have a mixture of guided activities and independent choices. I was impressed both by the quality of discussion amongst the children and the way in which they handled this change. It had implications for other areas of the curriculum and the class topic was certainly enhanced. Rationalizing what I have discovered so far has changed the way in which I will approach my teaching of technology, and will certainly affect the planning and imminent writing of the policy statement. It will also affect the organization of the classroom.

First, I felt some changes needed to be made in the location of some materials. Many items were freely available to the children and located in easily accessible boxes or trays that were all clearly labelled. Sheets of card, foils, transparent papers, and balsa wood were kept in the classroom but out of reach to the children. Some items were accessible to the children but only to be used with adult supervision, for example, the glue and ready mixed paints. I now felt that the children needed to see the variety available to them, and be encouraged to take responsibility for their use.

My attention was next drawn towards planning. I wanted to ensure that these experiences could be continued and developed. Other members of the staff were aware of the changes I had made. We identified opportunities for all of us to work together on the policy statement, and detailed planning of our rolling programme, in ways to extend pupil choice and develop the processes of choosing. It no longer seemed sufficient to ensure that the basic skills were taught and the tasks became increasingly 'difficult'. I wanted to have a more flexible approach where the children were developing their own ideas, making decisions and testing their options. It also seemed to me that we should have focused tasks each year to maintain the progression of skills and introduce the various tools available in the school.

FURTHER READING

Department for Education (DfE) (1995) *Design and Technology in the National Curriculum*, London: HMSO.
Dunn, S. and Larson, R. (1990) *Design Technology: Children's Engineering*, London: Falmer Press.
Futton, J. (1992) *Materials in Design and Technology*, London: Design Council.
National Curriculum Council (1993) *Technology in the National Curriculum*, York: NCC.
Tickle, L. (1987) 'Black spiders: art teaching in primary and middle schools', in L. Tickle (ed.) *The Arts in Education: Some Research Studies*, London: Croom Helm.
Tickle, L. (1993) 'In search of quality in Middle School curriculum', in T. Dickenson (ed.) *Readings in Middle School Curriculum*, Reston, Virginia: National Middle Schools Association.

Chapter 14

Measuring success

John Seaward

> At the heart of the matter is the design process. This is the process of problem-solving which begins with a detailed preliminary identification of a problem and a diagnosis of needs that have to be met by a solution, and goes through a series of stages in which various solutions are conceived, explored and evaluated until an optimum answer is found that appears to satisfy the necessary criteria as fully as possible within the limits and opportunities available.
>
> (Eggleston 1992: 26)

Like many teachers of primary design and technology who saw its introduction into the primary school of the early 1980's, I began by introducing technology into the curriculum using a problem-solving, design-process approach. Books such as *Problem Solving in School Science* (Johnsey 1986) advocated that approach and placed an emphasis on the 'process' in aspects of the curriculum which evolved into design and technology.

As primary technology developed, the introduction of Programmes of Study and Attainment Targets reflected the principles of the design process. The idea that a finished product should be achieved by following a sequence of steps seemed to gather momentum. Given that technology was a 'new' subject, this notion that an easily understood set of steps could be taught appealed to many primary school teachers, including myself.

Various models were developed which sought to identify the key activities of 'the design process' and attempts were made to explain the relationships that existed between them. Simple linear sequences evolved into more complex models such as 'interacting design loops' (Kimbell 1987). Using these models as a starting point for my own teaching, I became involved in what some have termed 'problem-based learning' in which, 'the actual process of solving the problem may be unimportant' (Hennessy and McCormick 1994: 94). The purpose of the activity is to help pupils understand certain concepts or ideas.

Moving into technology from an experience of primary science, this approach seemed to offer a way into a new and sometimes confusing subject. However, as my interest in primary technology developed and my knowledge and confidence in the subject grew, I began to move towards teaching 'problem-solving methods', which Hennessy and McCormick described as being when: 'the processes involved in solving the problem are the focus, and understanding of concepts (conceptual knowledge) is usually of secondary importance' (Hennessy and McCormick 1994: 94). So as the emphasis of my teaching moved towards teaching problem-solving methods, the organization of the children's activities changed. With this move towards teaching problem-solving methods, I soon saw the importance of the teacher developing a clear understanding of the sub-processes involved in problem solving. This chapter is based on a study of my attempt to develop an understanding of how one of those sub-processes, evaluation, is used by the children in my Year 5/6 technology group.

When I looked carefully at a number of design process models, ranging from the earlier simple linear type through to more complex interactions, it became clear that although the interaction of the various sub-processes changed from model to model, the sub-processes themselves remained common to a large number of models. The differences seemed to be about their interaction with each other rather than whether they should be included in the model. Evaluation appeared as one of the key sub-processes in all the models I investigated. I also became aware that evaluation was reflected not only in the design and technology curriculum, but appeared as an important component in cognitive development theory more generally.

This is highlighted by Jerome Kagan in a description of problem solving when he discusses five processes found in cognitive functions. He articulated the idea of an 'executive process' emerging around a child's fifth year:

> 'The fourth process in problem-solving evaluation, pertains to the degree to which the child pauses to consider and assess the quality of his thinking. This process influences the entire spectrum of mental work: his initial perception, recall, and hypothesis generation.
> (Kagan, Pearson and Welch 1966: 305)

With the introduction of technology in the national curriculum, evaluation was initially one of the attainment targets:

> Pupils should be able to develop, communicate and act upon an evaluation of the processes, products and effects of their design and technological activities and those of others, including those from other times and cultures.
> (DES 1990)

The revised Orders of January 1995 also placed evaluation as a key component appearing in the Programmes of Study, explicitly in designing skills when:

> Pupils should be taught to evaluate their design ideas as these develop, bearing in mind the users and the purpose for which the product is intended, and indicate ways of improving their ideas.
>
> (DfE 1995)

And also in making skills when:

> Pupils should be taught to evaluate their products, identifying strengths and weaknesses, and carrying out appropriate tests, e.g. on strength, user reaction, function.
>
> (DfE 1995: 5 para. f)

Key Stage Two level descriptions relating to evaluation appear in relation to Attainment Target 1 (AT1): Designing, in the following forms:

Level 2 'They reflect on their own ideas and suggest improvements.'
Level 3 'They make realistic suggestions about how they can achieve their intentions and suggest more ideas when asked.'
Level 4 'They evaluate their work as it develops, bearing in mind the purpose for which it is intended.'
Level 5 'Pupils evaluate ideas, showing understanding of the situations in which their designs will have to function, and awareness of resources as a constraint.'

(DfE 1995: 14)

In relation to Attainment Target 2 (AT2): Making, they also apply to pupil performance through the Key Stage:

Level 2 'Make judgements about the outcomes of their work.'
Level 3 'Their products are similar to their original intentions and, where changes have been made, they are identified.'
Level 4 'They identify what is, and what is not, working well in their products.'
Level 5 'They evaluate their products by comparing them with their design intentions and suggest ways of improving them.'

(DfE 1995: 15)

In the light of the evidence of the importance of evaluation in the national curriculum, its place in the design process, and its significance in cognitive development theory, I began to question the evaluation capabilities of my Year 5/6 technology group. At the start of the academic year, I made some general observations on the evaluation capabilities which I could detect. When I carried out a brief review of the children's performance in design and make activities, specifically in their ability to use evaluation effectively, it appeared that some children were operating at a much higher level than others. These were just impressions. Understanding how the children used evaluation in their design and made activities, how effective their evaluations were, and how they regarded the opportunities to evaluate their own work, became the starting point for this classroom based research. I wanted more than just impressions. I hoped that a better understanding of the children's capability in using evaluation in their design and make tasks could be used to achieve better quality of outcomes, by finding ways of teaching 'evaluation' more effectively.

This study focused on a class of twenty-nine, fourteen in Year 5, fourteen in Year 6 and one in Year 7. Two of the children have a statement of special educational needs and a third is in the process of being assessed. There are fourteen girls in the class and fifteen boys. Design and technology was introduced into the curriculum in 1991 and is part of an overall curriculum 'map' of the various elements of the national curriculum. It has its own designated time of one hour per week. However, there is considerable integration with other subject areas, in particular art and science. The dedicated lessons are used for focused practical tasks, when skills and knowledge from the Programmes of Study, are practised and acquired. Every half-term the children are set a design and make task, which is used as a basis for teacher assessment. I began my study by gathering documentary evidence, using the children's assessment records.

In considering the teacher assessment records, I decided to concentrate the analysis on the assessment of the group when working on the Programme of Study relating to the children being taught to: 'evaluate their ideas as these develop, bearing in mind the users and the purposes for which the product is intended, and indicate ways of improving their ideas' (DfE 1995: 4 para. g).

Using the level descriptions relating to evaluation, judgements could be made about the children's levels of attainment. Of the group of twenty-nine children, ten were assessed as achieving AT1 level 4, that is they were able to 'evaluate their work as it develops, bearing in mind the purpose for which it is intended'. Twelve were assessed as working towards AT1/4, and seven were assessed as working towards achieving AT1/2. That is, there was still not enough evidence that they could 'reflect on their own ideas and suggest improvements'. From the results obtained from the documentary evidence, I decided to focus on the group of ten children who had been assessed as working as AT1/4, and the group of seven children who had been assessed as working towards AT1/2.

The next step was to observe the children in two groups, working on three design and make tasks, which were presented to the children in the form of problem-solving activities, and reflected work on skills, knowledge and investigations carried out in both the design and technology and science curricula, during the proceeding half-term. Each of the tasks used activities devised by Johnsey (1986), all based on paper structures. The observed groups comprised three children who had been assessed as reaching level 4 and three children assessed as working towards level 2.

The children could draw upon previous experience of using paper materials in construction work. By using familiar materials I felt that the children could move through the various sub-processes and this would allow some degree of success and a greater opportunity to observe evaluations taking place. A time of two hours was allocated for the completion of each task, including a final evaluation by the children. Each child received a work sheet outlining the design brief and giving advice on the location of equipment, materials and research resources. Time was spent at the beginning of each activity ensuring that all the children understood the vocabulary and key ideas contained on the worksheet. During each of the design and make tasks the children worked individually, sharing a working location and resources.

Assistance was given to those children with learning difficulties, especially with their application of skills, but I tried not to unduly influence any of the decisions made by the children or offer any form of help with the evaluation of their products. The three design and make tests were:

- Using only a sheet of A4 paper, make a tower that can support a 1kg mass 21cm above the floor surface. There are many ways this can be done, so discover and record as many as you can (Johnsey 1986: 55).
- Design and make a container to hold a single egg using only one A4 sheet of card. The container should protect the egg against pressure or impact from all sides (Johnsey 1986: 56).
- Make a structure that holds a marble out as far as possible from the edge of your desk. You may use only four sheets of A4 paper, one sheet of card, 15cm by 15cm, and glue or sticky tape (Johnsey 1986: 61).

For each of the three activities the two groups were observed over the period of time taken to complete the tasks. One group was observed directly by the teacher, the other being videotaped and the film watched at a later date. Every effort was made to make the video equipment as unobtrusive as possible. This was relatively easy as the children are familiar with equipment in the classroom. The data from both the videotape and the first hand observation was recorded on a response sheet.

The groups worked in a small area, so that they could be closely observed and filmed in the way they carried out their design and make tasks. I decided to concentrate on three aspects of the way they worked:

- the time spent on analysing the problem, research, making and evaluating;
- the number of attempts at a solution;
- the level of success when assessed against the design brief.

Evidence from the observations seemed to show that the children assessed at AT1/4 spent a much longer period of time in the design stage. They spent a longer proportion of their time reading the brief, defining the problems and considering solutions. Time was allowed for the evaluation of ideas. Although overall this group spent the majority of their time on the actual making of their products, this was less than the time spent by the other group. When their success rate was studied it was found that each had made fewer attempts to arrive at solutions, and most of them had been successful. They had a clear understanding of the design brief and used it in full as the basis of their final evaluations.

The children assessed as working towards level two moved very quickly into the 'making' stage. They spent less time seeking to understand the problem, which had been set out in the design brief, and less time considering potential solutions. They made a greater number of attempts compared with the children at AT1/4, with a correspondingly higher failure rate. They appeared to spend very little of their time evaluating or making improvements, often picking one solution with only the barest consideration of the initial brief. Evaluation often appeared only when their making had been completed. In contrast the children working from the full design brief appeared to be constantly evaluating their ideas whenever opportunities arose, and did not see the evaluation as something that came only after 'making' had been completed.

It might be important to note that throughout the school there has been an emphasis placed on the importance of self evaluation, across the whole of the curriculum, over the past three years. The children are constantly reminded of the need to check their work carefully and to see this as an ongoing process.

When the group of children who were judged to be working at AT1/4 were identified to the class teacher, it was confirmed that they also worked at a high level in other subjects, notably in language, maths and science.

After completing their design and make tasks the children were asked to complete a questionnaire. The questionnaire sought to provide feedback on the children's attitudes towards the self evaluation of their design and make task.

The analysis of the pupils' responses showed a difference between some of the children's perception of the success of their completed products, and the teacher evaluation of the same product, in terms of its match to the brief. For task one, 65% said their product matched the original design brief, whereas I judged only 24% had made a product that matched it. The responses for the other two tasks showed a similar pattern of a higher number of children saying yes and the teacher assessment showing a much lower number.

This difference in the assessment of the tasks could be explained by the criteria used by the children and teacher in evaluating the finished

Table 14.1 Feedback on self evaluation

	Task 1		Task 2		Task 3	
	Yes %	No %	Yes %	No %	Yes %	No %
Could the children repeat the full design brief?	31	69	34	66	31	62
Did the finished product match the design brief?						
Children's evaluation	65	35	62	33	66	34
Teacher evaluation	24	76	33	62	33	62
Did the children find the evaluation easy?	69	31	72	28	66	34
Did they seek help from another child in carrying out their evaluation?	17	83	23	72	62	38
Did they like carrying out their own evaluation?	76	24	66	34	69	31
Did they have any worries about carrying out their own evaluation?	21	79	17	83	21	79
Did they think that carrying out their own evaluation was important?	90	10	100	0	100	0
Did they like sharing the results of their own evaluation with the whole class?	59	41	17	83	21	79

product. Both the children and teacher were asked to evaluate the products against the original design brief. The brief was given to the children in the form of a short written statement and time allocated so it could be discussed and any questions answered. The children were asked at the end of the tasks to repeat the original briefs. When asked to do this for the first task 31% could repeat the main criteria of the original brief; 69% did not repeat the full brief, they repeated only part of it. These children appeared to evaluate their product against only part of the brief and judged their work successful against that. The teacher evaluation was carried out using the full original brief. The same pattern repeated itself in the two other tasks.

Perhaps this was a function of the way I had set the tasks and the process of evaluation. Hennessy and McCormick state that:

> Although the way in which tasks are presented and adapted to pupils' needs and the degree of explicitness about the values of the task are known to be crucial factors in performance (Black 1990; Roazzi and

Bryant 1992), we found that teachers frequently did not initially make clear their aims and assessment criteria.

(1994: 101)

One of the children offered the comment, when reflecting on their activity, that I had not explained it very well. The design brief was presented to all the children in the same manner. They all had the opportunity to clarify with me any aspects they did not understand. If my introduction to the task was not clear, why did one group of children appear to have little trouble in carrying out their evaluations successfully against the criteria of matching the brief? It was clear that despite my care in presentation some children did not understand the design brief fully, to a similar extent in all three tasks.

From my discussions with the children and their responses to the questionnaire, I inferred that many of the children appeared to see the main aim (the task itself) and the sub-aim (the qualifying condition) in the original briefs. They then appeared to prioritize, taking the first aim and disregarding the sub-aim. This view was supported in the observations of their work in carrying out the design tasks. There are clear implications here for the presentation and organization of the tasks, and for the presentation of evaluation activities themselves.

When asked if they found evaluation easy, a pattern was repeated in the children's responses across the three tasks, with over two thirds thinking it was. When asked if they thought that self evaluation was important, the children responded overwhelmingly in the affirmative in relation to all three tasks.

Studies by Kagan (cited by Mussen, Conger and Kagan 1974), highlighted the differences in the ability of children to evaluate in problem-solving activities. Kagan offers a description of the differences in the way children evaluate:

> Some children accept and report the first hypothesis they produce and act upon it with only the barest consideration for its appropriateness or accuracy; these children are called impulsive. Other children devote a longer period of time to consider the merits of their hypotheses and to censor poor hypotheses; they are called reflective.
>
> (Mussen, Conger and Kagan 1974: 305)

The description of a reflective child seems to fit the group of children operating at a higher level of attainment. When observed working on tasks, they appeared to spend a longer period of time before acting on a solution. Once decided, they persisted with the task and made fewer errors. The description 'impulsive' is closer to the working practices of the children assessed as working at a lower level.

Apart from matching children to the profile of impulsive and reflective 'types', three tests of reflection–impulsivity have been used by Kagan:

Matching Familiar Figures, Picture Completion Test, and Extrapolation Reasoning Test. These were used so that the prediction that impulsive children would make more errors than reflective children could be tested. The results appear to support this prediction (Kagan, Pearson and Welch 1966). He claims that there is some evidence that a difference between reflective and impulsive children emerges as early as two years old. But there is also evidence that children do become more reflective as they mature or are placed in situations when they are working with more reflective children or adults.

I found the notion of reflective/impulsive children a possible explanation as to why the two groups performed differently when assessed on their evaluation capabilities. However, the tests used by Kagan to identify reflective/impulsive children do not appear to be available. If the children in the two groups are to be described in such terms, I would need to establish more clearly how each of them might be matched against the characteristics of reflective/impulsive conduct. I could endeavour, then, to devise strategies to encourage the development of reflectiveness in all of the class in ways appropriate to each pupil. The assessment of the class against the level descriptions showed a wide variance in the ability of the children, as judged in their finished products. There is also evidence, from the observation of the three tasks, that those children assessed as achieving the higher levels of product success were using evaluation in the problem-solving process. The group of children assessed as producing work at the lower levels appeared to show little evidence that they were using evaluation successfully during their design and make activities.

The identification of reflective/impulsive children could have implications for the organization of groups during technology activities. Mussen, Conger and Kagan (1974) suggest that impulsive children can have their behaviour modified and that under certain conditions they will begin to display the characteristics of a reflective child. This can be achieved in three ways, they assert: an impulsive child working closely with other children who behave reflectively; simply telling children to inhibit their impulsive responses to evaluating; and by the teacher displaying a tendency towards a reflective approach. Could these approaches be adopted into the teaching and learning strategies of the focused practical lessons and design and make tasks?

The grouping of children so that impulsive children are paired with reflective children in group work or a reflective child acting as a mentor, is a strategy that could be employed, provided they can be identified for pairing. That has its own problems, in the sense of its effects upon self esteem. For example, most of the children showed little anxiety when asked to engage in evaluation activities. They did however show concern when asked to share the results of their evaluations with the class. Some children felt that they would have their mistakes displayed in public. Yet most of the children appeared to have a positive attitude towards evaluation of their work. It would be necessary to make sensitive pairings or groupings, and agree the procedures for 'sharing' to encourage positive 'feedback' as an aid to the progress of each child's learning.

Most of the children answered in a positive way when asked on the questionnaire how they regarded the importance of evaluating their work, but they placed a far greater emphasis on the making of their products than planning or evaluating. This was also perceived by Hennessy and McCormick (1994) when they observed novice problem solvers. They comment that this is a result of both the children's lack of knowledge and the way teachers plan the activities. The explanation and instruction given to the children at the beginning of an activity could be structured to give support to children in each attainment level. Also the way the tasks are formulated could be reviewed to ensure that they allow for differences between children. The present practice of having one task, which the whole class engage in, could be replaced by using a design brief which could be modified for different ability groups, with perhaps a greater emphasis on the sub-process of evaluation for some groups.

The importance of the teacher acting as an 'expert' should not be overlooked, either. The children should have the opportunity to observe the teacher working on design and make tasks that contain a number of evaluation opportunities. This method is one that is used in many learning schemes, with the expert being observed by a novice, as skills are modelled.

I also need to expose the value, in a suitably clear way, of the processes involved in problem-solving activities. As I stated in my introduction there are numerous models that seek to describe the design process. But I need to find ways of helping the children to appreciate the nature of that process, so that they can appreciate the nature and place of the sub-processes.

There has been a move towards the use of manufactured products in primary technology. With an existing product the focus could be directed away from 'making' and more towards evaluation. Children could be encouraged to use the evidence available to explore objects and develop hypotheses. The testing of hypotheses will lead to evaluation opportunities. The range of objects is vast, with scope for the use of different materials, and the opportunities to investigate the strange and unfamiliar. However, if the needs of all the children are to be met, then the strategies that enable effective evaluation to take place must be made available to the children in relation to their own technology products too. Perhaps a link can be found between evaluation of others' work and self evaluation.

As a result of these observations and my thinking around the issues, I concluded that development and action to improve the evaluation capabilities of the children might be:

- Identify those children who are either effective or ineffective at using evaluation in their problem-solving activities as early as possible.
- Pair children so those who are judged to be capable at using evaluation are grouped with children who are assessed as being ineffective.
- Focus activities more overtly on the sub-process of evaluation within the design and make tasks, and the criteria on which their effectiveness will be judged.
- Help children to develop check-lists of evaluation criteria, to be used in investigations of existing products.
- Highlight the advantages of reflective working, encouraging a more deliberate approach which considers more than the first hypotheses that occur.
- Design and make tasks could be varied for particular groups, providing the opportunity to work at, and extend different levels of attainment.
- Demonstrate and model problem-solving activities for children to observe and analyse.
- More teacher intervention, working in partnership with a child or group of children.
- Checking that children understand design briefs, and appreciate the opportunity to reflect, research, create hypotheses and evaluate work.

FURTHER READING

Assessment of Performance Unit (1987) *Design and Technology Activity: A Framework for Assessment*, London: APU/DES.

Department of Education and Science (DES) (1990) *Technology in the National Curriculum*, London: HMSO.

Department for Education (DfE) (1995) *Design and Technology in the National Curriculum*, London: HMSO.

Eggleston, J. (1994) 'What is design and technology education?', in F.Banks (ed.) *Teaching Technology*, London: The Open University.

Hennessy, S and McCormick, R. (1994) 'The general problem-solving process in technology education: Myth or reality?' in F. Banks (ed.) *Teaching Technology*, London: The Open University.

Johnsey, R. (1986) *Problem Solving in School Science*, London: Macdonald.

Kagan, J., Pearson, L. and Welch, L. (1966) 'Conceptual impulsivity and inductive reasoning', in *Child Development* 37: 583–94.

Kimbell, R. (ed.) (1987) *Craft, Design and Technology*, London: Thames/Hutchinson.

Mussen, P., Conger, J. and Kagan, J. (1974) *Child Development and Personality*, New York: Harper & Row.

National Curriculum Council (1993) *Technology Programmes Of Study And Attainment Targets: Recommendations Of The National Curriculum Council*, London: NCC.

Chapter 15

Food and design technology
Where do we start?

Val Simpson

I developed a programme of work for Year 7 pupils in my 8–12 Middle (deemed primary) School, and although I felt that the pupils were acquiring useful skills, I was aware that I was not giving them the opportunity to develop their own ideas fully. The style of teaching I have adopted is one where, having set the context of the activity, I encourage the pupils to find their own solutions. I identify the technological knowledge that I want the pupils to understand and introduce the skills to be used in the project. As the project develops more skills are built in as the need arises. This is done by what I call 'spot-skill' teaching. In this way I draw the pupils' attention to a particular skill that needs either reinforcing or introducing, by demonstrating to the whole group. I feel that this has strengthened my teaching, giving me the confidence to let the pupils develop their own ideas within a given context/design task. At the same time I know where the teaching of skills is required.

To give the pupils the opportunity to work with food, I devised a series of food technology lessons, intending to adopt the same teaching style. I realized I had abandoned my preferred teaching style in the effort to implement the national curriculum as I understood it at the time. I had not identified any skills or processes that I thought the pupils would need and gave them freedom to respond to tasks in ways which focus more on the implementation of their own ideas.

I felt that the situation was too difficult to cope with because I spent far too much time moving from one pupil to another and teaching the skills that the individual required in order to pursue his or her ideas. I was also aware that the pupils were working in a potentially hazardous environment, using tools and equipment that were unfamiliar to them. I had tried very hard to get away from the 'Blue Peter' approach, but had succeeded in going from one extreme to another!

I decided, therefore, to employ a variety of teaching strategies, to investigate where the teaching of skills was most useful, and to see if there was any evidence to suggest that some methods were more restrictive for the pupils in terms of their opportunities to solve tasks creatively and individually. The latter was judged by the nature of the end products.

The design and technology programme was linked to work in R.E., and comparisons between Hinduism, Judaism, Sikhism and Christianity. The designated design and technology sessions used food as a material and I taught all the children in Year 7 on a rotating-group basis, in a designated technology classroom. A list of basic skills and learning opportunities which we hope the children will acquire and benefit from guided my decisions:

- Planning
- Making choices in designing the products
- Organizing time
- Preparation, using a range of equipment/tools in the preparation of food cutting (knives, food processor); cleaning; chopping; mixing
- Cooking (baking, frying, boiling microwaving)
- Following written instructions and report writing
- Presentation (cutting and shaping; making choices in the selection of materials and forms for the appropriate effect)
- Applying the design process
- Decision making and cooperation (to work both on their own, and in pairs, small groups and larger groups)

I used a range of teaching methods and organizational arrangements, according to what I deemed necessary to develop each pupil's level of skill and acquisition of knowledge and experience. These activities included:

Information giving	Testing of products and recording results
Demonstration	
Guided questioning	Problem posing
Perusing sources for ideas	Instruction
Direction of formal safety requirements	Practical experience with tools and equipment
	Group discussion

A series of lessons was devised, and the plans recorded, as evidence of my intention to vary the instruction in 'skills' and the 'making choices' elements. Data were collected from each lesson in the form of pupils' written reports, evaluation sheets, audio-tape recordings and informal discussion. After each lesson I produced a report which acted as a record of my immediate reflections on events. The thoughts which follow were drawn from summaries of these extensive data.

Lesson 1

Getting to know the working environment
Health and safety

Skills and use of basic equipment in making coleslaw
Equipment: knives, graters, food processor

The pupils worked in small groups to prepare one ingredient and use a given piece of equipment. The foods were assembled to make coleslaw. The use of equipment and the finished product were then evaluated. This lesson was about teaching skills of vegetable preparation in an instructional way. The outcome would be the same for each group although there would be some variations in colour, texture and amount produced.

Lesson 2

Using materials provided, each child was to make a sandwich for a particular celebration that they had identified. The choices offered – any three fillings from the range available – were to help the pupils to make decisions about which of the foods would best suit their purpose. Individual presentation and whole group evaluation followed. The aim of this lesson was to give the pupils the chance to put some of their skills into practice, but to be offered some choice or resources, unlike the previous lesson, although the outcome was still product led.

Lesson 3

We made a variety of breads within the group following the idea that 'bread is not only a staple food in many cultures, it also plays an important part in many celebrations'. The children worked in small groups to make one type of bread which was compared with others in terms of taste, texture and ease of making. The aim of this lesson was to give the pupils the opportunity to follow a recipe and to become familiar with using the oven. In most cases the pupils organized themselves in groups of three or four and had to organize the tasks between themselves.

Lesson 4

A variety of books were available from the project lessons. The children were set the task: 'Using the resources available, look for ideas from other cultures which will help you to make a food for a celebration.' During these adapted lessons I asked the children to research foods from other countries and how they were used on festive occasions.

Lesson 5

Many foods are taken to the places of worship and distributed so I took the idea of wrapping food in pastry or pancakes and having either a sweet or savoury filling. I demonstrated both of these using an apple filling and then a meat filling. These were the basic foods to be used, but the children had to add their own ideas to make it more appropriate for a celebration from another culture.

Lesson 6

I gave the groups freedom to choose what they would make. The only stipulation I set was that it had to be suitable for a celebration that they identified and that they were responsible for deciding upon and providing the ingredients. The pupils were told that they had to plan their work at home and to make sure that they were clear about what they wanted to do. I asked for a time plan, a list of ingredients and materials to be provided, together with a plan of action. I pointed out several recipe books that they might find helpful.

The implication of these adaptations was to employ instructional teaching methods in the early stages through product-led tasks, followed later by open-ended tasks where the skills taught were dictated by the individual's needs. This was the first time I had used food with the groups, so in employing these strategies I was conscious that:

- I had no prior knowledge of the groups' capabilities, or that of individual children;
- it gave me a chance to assess the groups, not only in the skills they brought with them, but in how they worked together;
- in the first lesson of any teaching situation the teacher has to establish some ground rules, particularly in terms of behaviour, so that an effective working relationship can be established;
- the pupils need to know what is expected of them in a new subject/situation;
- rules relating to health and safety needed to be established.

When planning the programme of work I found it (inevitably) easier to identify the skills that would be required when the tasks were product-led than when the tasks were more open-ended. In the product-led tasks the skills and processes for the product were clearly defined for the pupils and taught in an instructional way, using demonstration of the skills to clarify the points made. This method of teaching was used when making coleslaw and making bread, and in a modified lesson, making pancakes and pasties. I also found that product-led tasks were easier to resource as I was able to anticipate the equipment and tools that the pupils would need.

When given the opportunity to develop their own ideas around the theme of celebration many pupils used the skills and techniques of bread making and vegetable preparation previously learned. For example, Laura chose to make a dragon, shaped out of bread, a skill she had learned in a previous lesson, and to use vegetables and meat to decorate it to achieve the desired effect. Mark and Chris also used their bread making skills when designing a 'kite', and their vegetable preparation skills when decorating the kite. Sam had similar ideas, which her design showed, but she altered her plans, choosing to make gingerbread instead of bread. The reasons for this change was 'that I thought it would look good with the icing and because it would be easy to cut the shape out'.

The teaching of skills in the open-ended lessons was done on a more one-to-one basis, usually when I felt it was necessary to intervene and help a pupil who was struggling to achieve his or her aim, or when I was asked for specific help. I felt that the pupils had more difficulty in coping with the open-ended tasks because of the wide choice of outcomes that were available to them, even when the context, celebrations, had been specified. I felt that the pupils adopted their own strategies to cope, using their personal level of skills in these open-ended tasks. The use of skills learned in the earlier lessons was indicative of their drawing upon the personal resources of prior experience.

There was a significant number of pupils who chose to make something they had done either at home, 'with mum' or in other year groups. Tamsin, Laura, Sam and John were some of the pupils who chose to make sweets or biscuits using a method that was familiar to them. They each reported that they had made the product before and they felt that they could use the basic recipe and alter it to fit their design. This was a clear indication that the pupils were using the 'building block' method of learning, i.e. having identified a skill they could cope with, were building on their experience to a different end. Andrew and Alex also employed this strategy when deciding to make cakes. Alex brought a cookery book from home with a recipe for scones that he said he had used 'lots of times before'. I asked him which celebration he had identified when deciding to make the scones. He replied that he didn't know but his mum suggested that he make the scones because he knew how to do them. Andrew also brought in a recipe which he had used at home on many occasions. This latter example illustrates that the pupils were not sufficiently confident or did not know how to build on the skills that they had. They seemed happier to repeat a process that was familiar to them.

Bringing ready prepared food to the lesson was a strategy used by some pupils, usually a home-made or bought cake. These pupils had decided to use the ready made product as the basis for applying other skills. The cakes had been brought in so that they could be decorated. It was clear from observing the pupils that this skill was new to them and it was one where they needed the most help. The majority of the pupils had failed to appreciate the complexities of icing, which they had chosen as the basis for their work. This was one skill that I decided to demonstrate to the pupils who needed it.

172 Val Simpson

Figure 15.1 Individual responses to open-ended tasks: the planning stage

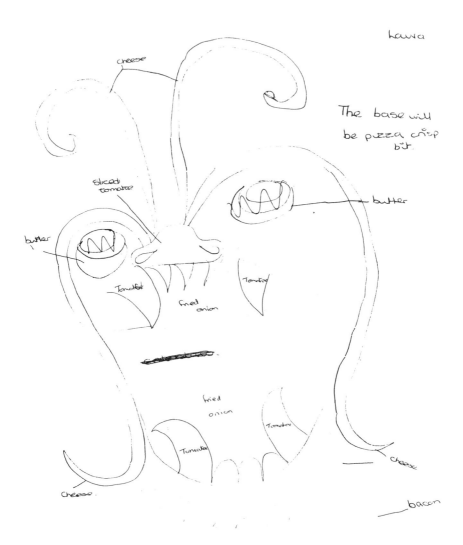

Ingreidiants

Bacon
processed cheese.
Tomatoes,
Mushrooms
Bread dough

What I must bring.

Tomatoes
Mushrooms
Bacon
Cheese
Tomatoe Pura.
carrot
grater

Time Table.

- 1·15 Roll out bread dough
- 1·25 Make shape (using template)
- 1·30 Frie mushrooms and bacon. and put on bread dough.
- 1·35 Heat cheese (melt it.
- 1·45 Pour on Bread Dough.
- 1·50 Gr Cook carrot.
- 2·00 Slice carrot put in dish.
- 2·05 Pour tomatoe puree on mouth etc
- 2·10 Put carrot in mouth.
- 2·15 Cut the left over cheese up. into shapes of teeth put on bread dough.
- 2·20 cut tomatoe up put on bread.
- 2·26 Put bacon and mushrooms in place.

This is my design

On this diagram the base of the car is ham, the number plate is cheese, the peoples heads are cream cheese, and the wheels are chocolate flakes, and that is all placed on the bread.

My Sandwich

When we were thinking and planning what we was going to do for our sandwich, we wrote down what materials and the equipment we thought that we would use. After we had done that we collected the materials we were going to use. Then we started, the first thing that I done was work-out how much time we had and then think out how much time I had got for each material. Then I cut the Ham into the shape I needed and layed it onto the chopping board, then I cut the cheese, and the cheese spread.

Afterwards I layed all the materials into place, but everything was soft, so I decided to toast the bread, when it was ready I put the materials back on to the bread.

That was when I got the chocolate flakes and layed them in place, and then I took a napkin and made a design with it and layed my open sandwich on to it, and then I had finished.

It turned out how I had designed it.

The one thing that I think I should have done was toast it at the end but I ran out of time.

The pupils had to justify bringing ready made food into the lesson and in most cases they were able to do that. Ann brought in a ready made cake which she spent the whole lesson cutting and shaping and decorating. She was able to justify this strategy because the aim of her chosen project was to learn how to cope with modelling icing and I had no doubt in my mind that she had been challenged by this new experience. Cindy responded similarly and her plan showed very clearly her reasoning for this decision to make the 'base' (fruit cake) at home, devoting lesson time to shaping and decorating a 'dragon' with coloured icing.

Following recipes was a typical response to the open-ended task and the pupils who employed this strategy fell into two groups. Those who brought a recipe of something they had made before, and those who brought in a recipe from home to try for the first time. The success of these pupils was varied, but in all cases there was evidence that a great deal of learning had taken place. Jacqui was very clear about what she learned: her 'failure' stemmed from the fact that she was ill-prepared by only bringing a list of ingredients to the lesson and forgetting to take note of the time and temperature that the cake required. Kerry and Sarah, on the other hand, appeared well prepared, but omitted to take note of the temperature that pastry needs to cook, resulting in disappointment. Paul also had problems because he hadn't fully understood the task he had set himself. He followed his recipe well but his failure to estimate the time taken for the jelly to set prevented him from completing the task on time. He had, however, decided to make biscuits to serve with his trifle and he was able to carry on with making those. Again he was following a recipe but he informed me that it was one that he had used before. The skill of following a recipe had been introduced to the pupils when bread making, but the success of being able to use and build on this skill was varied.

The way that the pupils responded to the open-ended tasks emphasized their ability to use skills they had already learned as a basis for a new situation, and the need to consolidate and extend them. Although I hadn't placed a strong emphasis on the tangible results the pupils appeared to do that for themselves, by choosing to make something that they were confident with. The way that Carl and Alex responded gave some indication of the parents' influence and involvement. I think that this parental influence was borne out in the way that Craig responded to the open-ended task too, when the cake ingredients were brought in with the suggestion that I would be able to show Craig how to make a cake. This implied that my job as the teacher was to demonstrate a skill, rather than to give the pupils the vehicle within the task to develop ideas.

Giving the pupils the opportunity to use the resources and materials effectively, but with a specified type of end-product, can lead to a variety of designs within the class. Opportunity to work within a set 'brief' or design task, and how they interpreted this brief, determined the range of outcome. In these cases the pupils were given a range of materials to choose from in order to plan what they would produce. The availability of information in the form of books, leaflets and teacher instruction, advice or guidance, gave the pupils something tangible to use to aid planning. Regardless of whether the lesson was product-led, or more open-ended, and regardless of whether the pupils were working individually, in pairs or in groups, at all stages in the designing and making processes, decisions had to be made about the characteristics of the object and the way to proceed. The pupils' ability to make decisions at every stage in the design process had an important bearing on the success, as well as the individuality, of the end-product. The opportunity to make decisions, though, the nature of them, was affected by the tasks, how they were set, the perception of the task, accompanying teaching strategies, prior experiences, available ideas and resources, opportunities to acquire new knowledge and skills, and methods of working 'on task'.

Taking account of food as an element within design and technology is still relatively new for me and I was aware that there were aspects which needed to be developed. The teaching of skills was a major issue. As confidence in my own ability and in the understanding of the 'Orders' grew, I have been able to shift the emphasis from just teaching basic skills for a predicted outcome, to teaching skills in the context of more open-ended task setting. I have tried to re-use this same pattern: the selection of basic skills in product-led activities steering the way towards giving the pupils more open-ended tasks which are 'context led' – i.e. defining situations and identifying design opportunities within them. As this happens the range of knowledge and skills needed by individuals increases and consequently so does their acquisition of them. I now need to develop strategies within the more open-ended lessons to ensure that the pupils get help when it is required by individuals or groups. However I am conscious that the skills that need to be developed are part, but by no means all of the Programmes of Study. Organizational, designing and evaluating skills need to be considered, and health and safety requirements extended. Perhaps most encouraging of all is the fact that the dilemmas involved have been identified, and that they can now be explored further in discussion with pupils.

FURTHER READING

Design Council (1987) *Design and Primary Education*, London: The Design Council.
Hammersley, H. and Hargreaves, A. (1983) *Curriculum Practice: Some Sociological Studies*, Lewes: Falmer Press.
Hitchcock, G. and Hughes, D. (1989) *Research and the Teacher: A Qualitative Introduction to School-based Research*, London: Routledge.

Chapter 16

'Seeing the light'

Nancy Wright

My lesson forecasts in design and technology take into account the fact that our school has a four-year rolling programme of cross-curricular themes: learning about jointed movement featured strongly in the year's planning. My outline plans for the start of the school year were, however, rather vague because I was going to be teaching younger children than previously experienced and I wanted to feel my way at the start.

Many of the children had either learning or behavioural difficulties and it did not take me long to realize that there were differences in their capabilities that appeared difficult to explain. It made me realize how much I had taken for granted certain levels of ability and understanding in my children of previous years. I therefore envisaged much rethinking of my expectations of these children. For this reason I chose to attempt to discover a way of developing my teaching, so that it would be as effective as possible in its contribution to their learning of design and technology, and so that I would be reassured that it was so.

In order for me to teach effectively, I hoped to be able to live up to the expectations of teacher-practice proposed in the DES Curriculum Matters 9:

> For most pupils, achievement through a theoretical or abstract engagement with a problem is incomplete without activity which puts their thinking into a framework that is real to them. Many need to start from the practical problem, without which a secure grasp of theory and abstraction is exceedingly difficult. They need to feel competent and productive and to derive satisfaction from 'getting things done', whether they are working from theory or through pragmatic, practical experimentation. These concerns are central to the best practice.
>
> (DES 1987: 8 para. 20)

The children's progression, I felt, depended upon their conceptual awareness and grasp of ideas, and their ability to make use of their learning in practical terms. It was my responsibility to find the most effective means of helping them to achieve and apply understanding. The most satisfying proof of the children's learning was when I could see, by the looks on their faces, that they had 'seen the light'. It was of immense importance to me to find out what caused this reaction and, importantly, to find whether the children themselves could recognize what caused their moments of understanding. This aspect of evaluation of my teaching/their learning was one that I wanted to develop with the children. I also wanted to find out how best to bridge what I perceived as a gap between the technological understanding of infants and juniors (i.e my new class and those I had previously taught).

When planning for the 'jointed movement' project, I chose to cater for the children's conceptual development in relation to that topic. I endeavoured to give them the right sort of tasks, the right sort of learning environment and the right sort of teaching styles compatible with their needs as I saw them, in relation to this particular aspect of their understanding. The key question in my mind was what constituted the 'right sort'. The chosen class-project themes were 'Underground' and 'Shops and Markets'. Jointed movement was used as the common technological factor within these themes.

I wanted to enable the children to investigate lateral, pneumatic, and rotary movement via a programme of activities based on the study of hinge joints, ball and socket joints, axles, bearings, sprockets and cogs. I did not plan to spend time on the related topic of activating these movements, except by hand, as time was to be spent on forms of energy later in the school year.

I aimed to provide the means whereby the children could develop an understanding of the appropriateness of particular joints for their intended use. My lesson-plans gave details of my teaching of concepts of jointed movement and also contained reference to work that I did as part of my own personal development. (This work was done with the intention of using it for INSET purposes within the school and local pyramid of schools, and provided an opportunity for me to analyse how I, the teacher, adapted to my own learning.) I made a model tipper truck, via trial-and-error design, incorporating pneumatic and electronic systems, and used the sort of materials and equipment that would be found in a well-equipped design and technology department. The children were given the opportunity to discuss and share my problems in the same way that they were encouraged to share their own problems.

My lesson-plans were developed in response to my judgement as to what would be effective in teaching hinge, ball and socket, and other rotary-movement joints. I was careful to leave room for adaptation or amendment as a result of the children's response to my teaching.

It was important for me to identify the range of learning opportunities that I gave to the children, and the following list contains situations that could be found in my classroom, according to circumstance. Part of my research was to find out from the children whether I needed to add to or prioritize any of the following:

- children discussing work together while working;
- teacher demonstration of skills or processes;
- design and making through trial and error;
- adult one-to-one assistance;
- children helping each other;
- questions that enhance understanding;
- blackboard information, as required;
- use of television, overheads, worksheets;
- linking a task to a real-life situation;
- experimenting and playing with materials;
- making a display for use as a teaching aid;
- recording of work: writing, drawing, speaking, photography;
- displays of work done, shown outside the confines of school;
- participating in activity on own ... with partner ... in group;
- evaluation of work after completion;
- review of work achieved over a period of time;
- praise given for effort as well as quality;
- selection of work for 'Good Work' Assembly;
- encouraging the children to ask own questions, do own research;
- teacher participation as supervisor for safety matters.

I realized that the children would have to be familiar with the contents of the above list, in the sense that they would be able to recognize whenever they met with any of them as features of particular lessons. My lesson-plans therefore incorporated identification of specific teaching methods and I hoped to obtain feedback from the children via their own evaluations. The children thus had the opportunity to share with me their views on their preferred learning opportunities, in the sense of these kinds of teaching techniques. This report describes the teaching projects in which I aimed at effective development of the children's understanding, at researching whether, and how, I achieved it, and at researching the children's preferred 'effective' approaches which helped them to 'see the light'.

LESSON PLAN ONE: FOLDING AND BENDING

Tasks Design and make an article that has movement caused by folding and bending; show direction of joints, such as in a 'bent leg'.

Activities Concertina-type pleating, experimenting with materials, presentation techniques, cutting and measuring.

Materials Paper (various kinds), card (various kinds), wood, tinfoil, plastic sheeting, fabrics, Sellotape, masking tape, PVA glue, scissors, snips, rulers, colouring media, skeleton model.

Skills Cutting, folding, measuring, sticking, thinking.

Teaching methods Demonstrate folding techniques. Display Jack-in-a-box, a salad roll, skeleton model. Work with small group then allow them to disseminate or share their knowledge/skills with partners. Rotation of activities. Evaluation after each session to encourage thought processes. Presentations of experimentations. One-to-one assistance where required.

Concepts What sort of movement is made possible by a straight fold? Can bones be folded? What allows the leg to bend at the knee? How does Jack stay in his box?

Observation Lots of 'seeing the light' during the experimenting with rigid materials, also when involved with the task 'leg-bend'.

LESSON PLAN TWO:
HINGE JOINTS EXPERIMENTING

Tasks Find things in and around school that have moving parts that open and close. Draw them and try to explain why they need to open and close, and also how, if you can.

Activities Close observation. Experimenting with materials. Sharing ideas. Evaluation.

Materials Paper, tinfoil, card, twigs, pea-sticks, PVA glue; rulers, scissors, Sellotape, manufactured hinges, hinged objects.

Skills Cutting, rolling, glueing, classifying, communicating.

Teaching methods Demonstrate rolling paper techniques. Allow trial and error in design and making. Praise all children for their efforts. Encourage positive evaluation of products. Children working as individuals, paired and grouped.

Concepts How strong do joints have to be to be effective, safe, reliable? What is the range of movement created by hinge joints? How does the hinge hold itself together?

Observation Some keen competition to find hinged objects that no-one else had found! Surprise expressed at the quantity. Lots of effort put into work. Very little input from me. Very busy, all of them. Searching for hinged objects continued voluntarily for rest of week!

LESSON PLAN THREE:
HINGE JOINTS APPLICATION

Tasks Using the materials provided, design and make a four-sided folding frame. Work in groups of no more than four. After half an hour, class evaluation to choose best design. Using best design, make the frame out of wood and manufactured hinges.

Activities Organization of materials. Organization of work. Making reasoned choices.

Materials High-density foam blocks, drawing-pins, paper, card, wood off-cuts, hinges, screws, pointed awl, screw-driver, mitre-block and saw, sandpaper.

Skills Ability to cooperate, thinking, cutting, folding, turning screws, using mitre-saw, sandpapering.

Teaching methods One-to-one assistance where required. Children working with and learning from each other. Demonstration of marking positions of hinges/holes, using awl, mitre-saw and sandpaper-block. (Older children supervised but unaided.) Giving children freedom to pursue own ideas.

Concepts Does it matter where you put the hinges? What effect does movement have on the hinge joints? How can this problem be overcome?

Observation Many 'saw the light' when they were making their foam frames. There was a keen interest in succeeding within the half an hour. Younger children had no idea that wood could be so hard. Loved the idea of a 'secrets box' made out of the off-cuts. Children want to put lids at each end, left them to puzzle out how. (Will help/give them a clue if necessary.) Liked the way the children paired off to put the wood-screws in so that everyone could be involved.

LESSON PLAN FOUR:
HYDRAULICS, PNEUMATICS

At this stage of our work, I thought that it was appropriate to introduce a way of opening and closing a gap so that the children could see it eventually applied on my tipper-truck model. My model was in the process of being made and there was enough ready for the children to have some idea of the design problems associated with the tipper movement.

Tasks Working in pairs, using any combination of the equipment provided, design a movement system that is capable of pushing and pulling. Record any discoveries that you make about what effects the way the system works.

Activities Working to own ideas and then testing, modify as required. Recording results as basis for further development.

Materials Syringes (various sizes), plastic tubing (various lengths, thicknesses), snips, T-joints, air, water.

Skills Attaching tubing to syringe nozzles. Applying knowledge to develop work. Measuring, reading scales. Discrimination as to what is useful to record.

Teaching methods Children experiment with syringes without tubing to start with. Children learn from each other. Providing motivation for them to think for themselves in order to help me.

Concepts That a series of connected joints can result in indirect movement. Can anyone see any connection between the way a hinge can be taken apart and the way the inside of the syringe travels?

Observation The 'grapevine' became fully active as soon as individual pairs discovered something new. Any messing about with water that was not connected to the task had to be discouraged very early on. Liked the way pairs of children played tricks on each other, lifting up things that were placed across gaps in tables from underneath. (Valuable activity; children now very much involved with tipper.)

LESSON PLAN FIVE: LEARNING METHODS

I thought that the time was ready for the children to revise what they had achieved in their learning. They were beginning to gel as a class and many appeared to be maturing in their approach to learning

Tasks Discuss with each other what you have learned; try to remember how you learned it. Choose someone in your group to make a note of all the different ways. All groups join together and share your findings. Let me have a list of what you have remembered.

Activities Responding appropriately to instructions, and evaluation of their own ways of learning.

Materials Samples of all kinds of work done to date with the exception of my tipper-truck.

Skills Communication using technological vocabulary; listening, while others speak on their behalf.

Learning methods (as identified by the children) Trial and error using own ideas. Step by step as shown by someone. Learning from each other's discussions. Look, touch, play and experiment. Being asked questions. Teacher telling us clues. Friends teaching us how to do it. Writing about what we do. Finding things outside that are like what we're doing.

Observation Pleased that the children are responding to a variety of methods. (What about the ways that have not been remembered?) Am reassured that the children's learning is identifiable and that from now on I can seek to find preferences amongst the children. There appears to be satisfaction within the class that all their contributions are represented.

LESSON PLAN SIX:
BALL AND SOCKET TECHNOLOGY

Tasks Go into the playground (supervised), where you will find my car and trailer. Have a good look at how the trailer is joined to the tow-bar and draw what you see. When you have finished, ask me to demonstrate how it works. Then find something in the classroom that uses the ball and socket method to make it move. Draw what you find and write about any differences you find about how they are made and how they move. (Try to work on your own for the last task.)

Activities Close observation, recording of findings by writing and drawing, looking for related objects.

Teaching methods Ask questions during demonstration that promote thinking, give children opportunity to apply thinking skills.

Concepts What happens to the trailer when the car goes round bends? Are the roads perfectly smooth? What are the problems for a car and trailer if they aren't? What happens when a car and trailer go over a hump-back bridge – down a dip? How does a ball and socket joint help to overcome the possible problems? What is the range of movement provided by ball and socket joints?

Observation Many children 'saw the light' when they watched me drive first to the left and then to the right and saw what happened to the trailer, then saw me stop in the dip, the car facing uphill and the trailer facing downhill. Several of the children thought that the air-freshener ball was not a proper joint because it came apart in their hands.

LESSON PLAN SEVEN:
BALL AND SOCKET BODY-JOINT

Tasks Look at the arms on the model skeleton to see what sorts of joints they have. Using the materials available, make an 'arm' of your own, complete with 'hand'. Present your arm on card so that it is part of a 'Person using their arm' display.

Activities Assembly of components, joints to include hinge and ball and socket. Cutting of plastic pipe as required, filing edges. Presentation by attaching to display card, drawing and colouring as appropriate.

Materials Various plastic tube containers, pipe to cut to size, table-tennis balls, wooden beads, large tapestry needles, thread, adhesive tape, card, colouring media, scissors, junior hacksaws, G-clamps, hand files, glass-paper, benchhooks. N.B. Any holes that were needed in the plastic were made by me when required.

Skills Cutting, sawing, threading, taping, filing, stitching, glasspapering, application of observation of arm and hand relationship (thinking needed).

Teaching methods Demonstration of cutting pipe, filing rough edges, smoothing with sandpaper. Children to work to own ideas, choosing own materials. Partnership work encouraged, one helping the other, then vice-versa. Advice given, to check each other's 'arms' in relation to 'hands'.

Concepts How does the arm hang in relation to the side of the body? Where is the thumb in relation to the hand/arm? Which way does the elbow bend? What happens when the arm is raised?

Observation Many instances where, in spite of all my advice, hands were assembled back-to-front. Children helped each other to remedy the situation and in so doing, 'saw the light'. Some good responses to this work, very busy, much creativity.

LESSON PLAN EIGHT: EVALUATION OF LEARNING

The work done so far had gone to plan, so I thought that it was time to ask the children if they would help me to evaluate the list of learning methods as devised in Lesson Plan Five.

Tasks Use the list of learning methods that you have done. Draw a star next to the method that helps you best to understand your work. Put two ticks by the side of your second-best way of learning. Put one tick next to your third-best way. Put a cross next to the ones you really dislike.

Observations The children had no difficulty filling in their sheets and I was a little surprised to see that they were taking great pains to be independent in their evaluation, i.e. not being influenced by what their friends had chosen. 'Why does "Being asked questions" come bottom of the list?' I asked. 'I don't like being asked because I like to hear what someone else says', replied one of the boys. The next on the list of 'Least-liked' ways was 'Recording work', but they did admit that they enjoyed having their work in a folder to look at, as a reminder, and were pleased when it was chosen for 'Good Work' Assembly. We rated the results: 4 points for a star, 2 for two ticks and 1 for one tick. The list below shows the points totals in order of merit according to the whole class:

Table 16.1 Learning methods and results of voting

Learning methods	Points
Discuss with friends	38
Link to things found outside	32
Look, touch, play and experiment	27
Friends teaching us how to do things	21
Step by step being shown	20
Trial and error	19
Teacher telling us clues	10
Writing about/Recording work	6
Asking questions to find out what we know	0

This is where I kept my promise to the children that I would give them the opportunity to work according to their favourite methods of learning and teaching not just as a class, but as groups within it. Thus group one star choices were: discussing with friends, linking to outside application, teacher giving us clues, and friends teaching us. These were combined and I organized work based on them. The remaining children comprised group two, who chose to work according to step by step instructions, trial and error using their own ideas, look, touch, play and experiment, and recording work by writing and/or drawing.

I hoped that both groups would produce some quality thinking and learning, and set tasks for two projects catering for both groups. The first project was for the purpose of continuing our work on the concepts of jointed movement. The second was for the purpose of allowing the children to organize their own work in order to complete their tasks. I then assessed the children's concept-learning by the use of self-devised worksheets. I noted my personal impressions of the capabilities of the two groups, and their orientations to practical activities.

Group one children all showed confidence and had practical ability. Two of the group were 7 year olds who were showing signs of high potential and also included in this group were the five oldest children in the class. The rest of the group (six 8 year olds) showed ability to work cooperatively with the others and also use language skills effectively. The five oldest children shared their skills without dominating.

Group two children showed a willingness to learn but all required much reassurance and one-to-one guidance. Six of the children still had behavioural problems and four children, although cooperative, had special needs relating to forms of communication. I thought that the differences between the two groups corresponded to their choices of learning methods and I began to look for signs of independent activity in group two children as a sign of development of confidence in their work.

LESSON PLAN NINE:
PROJECT ONE, SWIVEL CHAIR

Task I am now going to the supermarket to buy some food. When I get to the check-out, I notice that the person at the till is sitting on a chair that is fixed to the floor, but the seat turns round. Using the materials provided, could you design and make a model chair that behaves like the one that I have described? You are to do this task if you are in group one. You have two sessions in which to complete the task.

Activities Making use of marbles, lids and grooves in order to provide turning mechanism. Finding a way of supporting the mechanism so that it can function as a seat.

Materials Marbles, art-straws, pipe-cleaners, tin-cans, lolly-sticks, Plasticine, snips, plastic lids, cardboard tubing, variety of card, masking tape, PVA glue, display boards of 'joints', paper, scissors, manufactured bearings for observation.

Skills Using initiative, new experience of above materials.

Teaching methods   I shall be available when needed. (According to group one, I should be giving them clues.) I shall also arrange for the children to go, in their groups, to the office to observe the typist's chair.

Concepts   How easily does a marble move? What use are bearings? How can they be kept in place? How can they help things to rotate?

Observation   Lots of fruitful discussion among friends resulting in them helping each other via look, touch, play and experiment; they all 'saw the light' when one of them managed to rotate a large book on a very small lid. Construction of chair required lots of clues from me, but they enjoyed the challenge. Much learning took place by trial and error. The group realized this and commented on the fact! There were also comments that they needed to observe the typist's chair to find out how it was made, and that they would otherwise have had a lot of trouble doing it.

LESSON PLAN TEN: PROJECT TWO, CONVEYOR BELT

Task   I am now going to the supermarket to buy some food. When I reach the check-out, I notice that I have to place my purchases on a moving surface, which keeps stopping and starting. My food gets carried along. Can you design and make a mechanism that behaves in such a way? Use the materials that are in the box. You are to do this activity if you are in group two. You have two sessions in which to complete the task.

Activities   Assembly of parts, contained in such a manner that they convey a supportive belt. Presentation of the resultant mechanism as a check-out counter.

Materials   Cotton reels, matchsticks, ribbon, stiff card, PVA glue, corrugated card, cuboid boxes, rubber bands, old pencils, snips, paper, colouring media, cogs, hole punch, dowel rods.

Skills   Using initiative, cutting, measuring, thinking, following instructions, presentation of product. Written and drawn recording of work done.

Teaching methods   As per group two requirements, but only providing step-by-step workcards with construction kits if requested, in order that those who prefer to work by trial and error do not have the former method imposed upon them.

Concepts How can one rotating object affect another? What happens to the direction of turn from one to another? How can an object travel (be conveyed) while the counter remains fixed?

Observation Epsom printer noticed as having sprockets and paper with matching holes. Two children tried to imitate this using Teko cogs and holepunch and card. Two more children tried copying the mechanism on a bicycle; using chainlink from Capsela and finding cogs to fit. Trial and error very much in evidence. Simultaneously, some children were playing with the corrugated card and discovered that it fitted when put face-to-face. 'Look what we've found' soon turned into 'Look what we've done' as they made their own cogs by wrapping the corrugated card around tubes and used corrugated card strip as the conveyor belt itself. There was uncertainty as to how to assemble the tubes so that they rotated within their cuboid box framework: Step-by-step example provided by myself showed one way of overcoming the problem.

LESSON PLAN ELEVEN: PROJECT THREE, SHOPPING TROLLEY/HANDCART

Choose either Task 1 or Task 2.

Task 1 We have finished our shopping and have everything stacked in the trolley ready to go to the check-out. Design and make your shopping trolley, selecting from the equipment provided.

Task 2 We have been learning about the Tudors and Stuarts and their searches for new foods. Pretend that you are taking some foods to market in your handcart. Design and make your cart, selecting from the materials provided. You are to work as a team, each sharing in the task that you have chosen. You have two weeks in which to complete the task.

Equipment/materials Wire coat-hangers, florist wire, jumbo art-straws, PVA glue, wool, string, heavy card, card tubing, square dowel, dowel rods, snips, junior hacksaw, scissors, variety of castors (in sets) and a variety of wheels, including one that I had made using lolly sticks and card.

Observation: Task 1 Coat-hanger wire abandoned as attempts to make trolley frame failed, wool used to imitate mesh within heavy card frames, castors attached to card strip base using florist wire, cardboard tubes used for handle, thought given to shape of trolley to allow interlocking, test drive successful in carrying articles, steering good. Cooperative work, shared tasks, delegated tasks, much discussion and planning before starting. Children very pleased with results.

Task 2 Much rummaging in the materials box, trying to choose wheels. One child had got a library book with pictures of carts (which were most appropriate according to historical facts). Eventually chose wheel that I had made, deciding to make another like it. Half the group got together to make wheel, the other half made cart body. Arguments with the wheel-makers as to how to attach wheels to cart. Various ideas suggested, resulting in agreement to glue wheels to axle then glue axle to cart. The first part was successful, the wheels were stuck to the axle. The next part was near disaster until one of the children realized what would happen if the axle couldn't turn. After a rethink including playing with dowel rods and art-straws wheels were pulled off the original axle and reassembled so that they turned.

Too much glue used, wheels stuck to sides of cart so wouldn't turn until I used a sharp blade to cut through and free them. This group were very reliant upon my help and their model eventually drew the attention of the whole class. Much learning took place on a trial and error basis because I thought that there was sufficient confidence being shown to withstand a temporary set-back.

LESSON PLAN TWELVE: VISIT TO SUPERMARKET

Several parents and Governors accompanied my class and they were very supportive in helping to make sure that the children focused their attention on the activities devised by the supermarket staff.

Reason for visit (in relation to design and technology) To consolidate the children's learning experiences and enhance their linking of model-making to real-life.

Tasks To observe, ask relevant questions and respond appropriately to practical experiences offered by supermarket staff. (Many of the children commented that they had been to a supermarket before but hadn't noticed the variety of shopping trolleys. Neither had they taken much notice of the check-outs.)

Follow-up I asked my class to complete their Shops and Markets theme by writing to the supermarket to thank the staff for taking such an interest in what we were doing. The children also designed a poster each, upon the theme Sugar for Energy, showing sports being played. I wondered whether the work that we had done in three-dimensions would affect their two-dimensional representation.

Observations Much interest was shown by each group of children (as reported back to me by group leaders) in the working of the conveyor by button control. There was some disappointment when they were unable to investigate the electronics involved. Some valuable lessons on aspects of Safety in the Workplace were given by staff, which I thought was a bonus from our visit.

Comments (sugar for energy posters)   I was interested to find that the results were not as I had expected. Nine of the children had made a dramatic improvement but the others seemed to have problems and responded to the language aspect of the poster more readily than they did the action drawing. (This was in spite of the fact that preliminary work had been done with resources of television sports' footage and use of P.E. time to try to produce posed action shots.)

We were, however, heartened to find that most of the resultant posters were a big improvement compared to their first pictures that had been drawn to represent Sports Day and there was general pride felt by the children in their achievements. I feel that continued practice with figure drawing now will pay dividends, building on the confidence beginning to be felt.

LESSON PLAN THIRTEEN: ASSESSMENT OF LEARNING

We had now come to the end of planned design and technology activities related to learning the concepts of jointed movement. I wanted to find out what the children had achieved and by what methods. During a review of the content of each of the Lesson Plans 1–12, interest was shown in Lesson Plan 5, when additional learning methods were offered by some of the class. These were therefore included in this assessment activity.

Task   You will each have two worksheets to complete: These will remind you of work done during our Moving Joints activities. The first will give you another opportunity to vote for your favourite ways of learning. (You may have changed your mind since the last time you voted and you may have something else to add that you have just thought of.) The other will ask you to remember some of the things you have been learning about moving joints. Please do your best and if you want to write anything extra about the work that you have done, please use the backs of the sheets. I shall let you know the results of your votes so that we can discuss them in class.

Observations   The worksheets were greeted with enthusiasm by many of the children but there were some children who found difficulty in responding to the voting in the manner asked for. We overcame this problem by discussing their responses on a one-to-one basis and narrowing down the chosen methods to the number required.

LEARNING METHODS: VOTING SESSION 2

Points were awarded on the basis of 4 points for the favourite method, 2 points for the next choice and 1 point for third choice. (The oldest child suggested this so that 'favourites' could win more easily.) The learning methods received points as follows:

Table 16.2 Learning methods and results of voting

Learning method	Year 3	Year 4	Total
Discussing with friends	26	16	42
Friends helping me	11	18	29
Look, touch, play, experiment	22	2	24
Teacher giving me clues	12	9	21
Step by step instructions	13	6	19
Displaying/talking about work	15	0	15
Trial and error using own ideas	5	3	8
Writing and drawing to help me to remember work	0	5	5
Being given materials and equipment to choose from	1	4	5
Imitating things using substitute materials	3	1	4
Finding things in real-life that help me	2	1	3
Being asked questions to help me say what I think	0	0	0

Table 16.3 Summary of possible scores on worksheets

Item		Score
A	Hinge Joints	6
B	Socket Joints	4
C	Swivel Joints	2
D	Castors	3
E	Cogs	2
F	Tool Safety	2
G	Design minimum	2
H	Voting maximum	1
I	Review of Work minimum	1

WORKSHEET SCORES

The worksheet recorded evidence of learning on a range of specific items from the topic, with possible score allocated to each item.

The top basic score attainable was thus 23, but there was a possibility of gaining extra by giving additional information for two of the questions. I used the sheets as an assessment instrument which would measure recall of learning in a way which showed differentiation by outcome. In advance I selected the items on the basis that those children who scored 20 and above showed evidence of high standard of learning and recall. Those who scored between 19 and 13 (inc.) showed evidence of 'satisfactory standard'. Any who scored 12 or less were regarded as showing evidence of 'poor quality of learning'. Eleven children scored 20 or over; twelve children scored 13–19 and the child who scored below 13 responded to oral assessment as shown on the class results which follow.

Table 16.4 Recorded actual scores

Child	Scores A	B	C	D	Item E	F	G	H	I	Total
L.L.	6	3	2	2	2	2	2	1	4	24
J.Sa.	6	4	2	3	2	2	2	1	1	23
L.S.	5	3	2	3	2	2	2	1	2	22
J.R.	6	2	2	3	2	2	1	1	2	21
R.S.	6	4	2	3	2	1	1	1	1	21
C.Po.	6	4	2	1	2	2	2	1	1	21
BJ.HH.	6	3	2	1	2	2	2	1	2	21
P.De.	5	3	2	1	2	2	2	1	3	21
A.E.	6	3	2	1	2	2	2	1	1	20
C.Ph.	6	4	2	1	2	0	2	1	2	20
H.G.	6	4	2	1	2	0	1	1	3	20
J.C.	6	3	2	3	2	0	2	0	1	19
S.W.	6	4	2	1	2	1	1	0	2	19
M.D.	5	3	1	2	2	1	2	1	2	19
R.R.	5	3	2	1	2	1	2	0	3	19
S.A.	6	4	1	1	2	0	1	0	4	19
G.O.	6	4	2	1	2	0	1	1	1	18
C.G.	4	2	2	3	2	1	1	1	2	18
D.D.	5	3	1	1	1	1	2	0	4	18
P.Du.	5	3	2	3	2	2	0	1	0	18
L.Ab.	5	1	2	1	2	1	2	0	3	17
F.F.	3	3	2	1	2	1	1	1	2	16
A.H.	4	2	2	1	2	1	2	0	0	14
J.St.	3	0	1	1	2	1	1	0	3	12
L.At.	(6)	(4)	(1)	(2)	(2)	(0)	(1)	(1)	(1)*	(18)
C.B.	(5)	(2)	(2)	(2)	(2)	(2)	(2)	(1)	(1)*	(19)

*The latter two children were absent at the time of the written worksheet but on their return to school, responded to oral assessment.

A useful aspect of the written work proved to be in the response from the children when their worksheets were handed back to them. All the 'weak' areas were highlighted by the above breakdown and there was sufficient interest and motivation to spend time on reinforcement of learning.

This was achieved by utilizing the children's favourite methods, i.e., 'discussing with friends' and 'friends helping me', with a little of 'teacher giving me clues about what to do'.

SEEING THE LIGHT

The worksheets were an attempt to obtain some idea of the children's knowledge in a 'nutshell', and also provided a means whereby the children had a record of some of their design and technological activities. Their conceptual development had been in the 'doing' of tasks and I used my ongoing assessment based on observation to ascertain a more complete picture. The following quotes are from the children on the subject of their seeing the light:

AE   When we made a hinge I learned much more than I already knew when I worked on my own playing with them. The point when I realized what I was doing was when I copied one.

CPO   We drew round the hinges and screwed the screws in. And that was when I understood what I was doing, because if the hinges were not straight the object would not open and close properly.

HG   When we drew ourselves and drew arrows where hinge movement was, that made me realize what I was doing.

CPO   I made a hinge joint out of paper – it was like fingers locked up together. I put a twig down the middle so it wouldn't fall apart. That's when I understood what we were doing.

RS   When we made the arms that we tied onto the card it made me think about what I was doing.

DD   When teacher was talking about arms bending I didn't understand but when we did the real thing I understood it.

GO   We wanted to carry stuff on a handcart. We drew it first. Me and S worked wrong at first. When we got it right it was good. I would like to do it again. Me and S enjoyed it.

JSa   When we were doing the wheels they got stuck. It was then when I knew what we were doing.

JR   When I joined the bits I knew I had to do it so it would be safe when it moved.

MD   Having to explain my work in Good Work Assembly made me understand it.

The quotes highlight the value of 'doing', and refer to aspects of trial and error problem solving, looking, copying, evaluating, and explaining to others the function of moving joints. I allowed a few weeks to pass before sharing with the children the results of their recent voting. My class now included (temporarily) a group of six Year 5 children who had already had experience of working with me on the theme of Movement, so I took the opportunity of including these children in the final evaluation session.

I gave time for the children to look at and discuss their worksheets and shared examples of how particular learning methods helped achieve understanding of jointed movement. To do this children worked in small groups and each group included a Year 5 child. The children added to and amended/clarified learning methods as appropriate, making use of the blackboard and devised a system for voting.

The learning methods listed below were presented for voting. Each method was allocated an identifying letter of the alphabet and a consensus of opinion ruled that five methods should be chosen by all, points to be given in order of preference, i.e. 5 points for the favourite, down to 1 point for the final choice.

A Trial and error methods using own or other person's ideas.
B Step by step instructions, whether spoken, written, drawn or demonstrated.
C Learning from each other's discussions during designing, making or display.
D Look, touch, play and experiment with real-life objects.
E Being asked questions to help us to think more.
F Teacher giving us clues when we get stuck.
G Friends telling us how to do things, as long as the friend knows how.
H Writing or drawing to help us to remember what we've done; worksheets.
I Finding links with everyday activity/outside links (not special outings).
J Teacher helping me on my own.
K Going on special outings, reading books, looking at pictures, watching television.

This list was compared with the previous voting showing that there was a change of 'results' each time we carried out a poll survey. My thoughts concerning this formed the basis for the next part of my enquiry.

CHILDREN'S OWN LEARNING: FULFILLING THEIR NEEDS

In the Schools Council Working Paper 75, *Primary Practice, a Sequel to the Practical Curriculum*, I found a section entitled 'Curriculum as Process'. In this, there was advice that:

Table 16.5 Results of evaluation: learning methods (in order of points)

	Yr. 3	Yr. 4	Yr. 5	Total Y3/4	Total Y3/4/5
K	42	21	22	63	85
D	42	14	7	56	63
G	28	20	8	48	56
A	26	14	5	40	45
C	15	19	10	34	44
B	22	10	6	32	38
J	17	12	7	29	36
H	16	8	10	24	34
I	19	4	1	23	24
F	6	7	8	13	21
E	9	4	6	13	19

One is to consider what skills and qualities today's young people are likely to need now and in their adult lives. Do they include, for example, such qualities as:

1. Communication skills, through whatever medium is most appropriate to the message and the intended audience;
2. Study skills, the ability to work and learn independently, the ability to organise information through themes, formulae or concepts;
3. Ability to work from written technical instructions;
4. Ability to define problems and find solutions to them, to plan, and to make decisions;
5. Practical and technical skills;
6. Personal and social qualities including coping skills and adaptability, ability to work co-operatively with others, sensitivity, imagination and creativity, self-esteem, and a sense of moral values.

(Schools Council 1983: 29)

During the course of their concept-learning about Jointed Movement, I judged that the children developed some ability in many of these skills and qualities, and their achievements were largely due to their own use of a range of chosen learning methods. What became apparent as my investigation developed was their desire to work independently of me as a formal instructor. I was very much in demand initially, but as the children got used to the idea that it was 'OK' to work with friends, using other methods to achieve their learning, my role became that of the person to go to if the children had problems that they couldn't handle themselves. Their final list gave me some insight as to the nature of their preferred method of learning and its effectiveness as they saw it. I began to relate effectiveness in gaining these skills and qualities to those methods as I connected that 'evidence' to my observations and impressions of events, as follows.

Communication skills featured in several chosen methods. Their desire to talk with each other during practical work was strong and became the top favourite in the early stages. Once they became acclimatized to working using shared ideas or known skills, the older children wanted to become responsible for finding out things for the younger children, using books where possible. They saw how valuable pictures could be as learning aids, and began to take a pride in their presentation of written and drawn work. Their skills in describing how joints could be assembled or moved were developed through the practical experiences of looking, touching, playing and experimenting independently. This method became popular enough to move up to second place.

The children's ability to work independently grew from confidence gained from familiarity with a concept or skill. The value of giving the opportunity for experimentation became evident as they made discoveries which they wanted to share. With my younger children, I did not ask for 'end products' every time they had an exploration session, and most felt they had learned a lot just by handling materials or an object. I found myself thinking of John Holt's quote of David Hawkins in his book *How Children Learn*:

> Children are given materials and equipment – things – and are allowed to construct, test, probe and experiment without superimposed questions or instruction. I call this phase 'messing about'. . . . In some jargon, this kind of situation is called 'unstructured', which is misleading; some doubters call it 'chaotic', which it need never be.
>
> (1991: 219)

The children's ability to work from written technical instructions was not a major feature in our concept learning and was eventually superseded by trial and error learning. Written instructions were available with construction kits, but those with pictures or diagrams were the most popular. The children's favourite format was the step by step version.

Practical and technical skills in relation to concepts of jointed movement were initially taught by me, but the learning often came when the children helped each other to apply them. Friends always played a prominent part in the children's learning, but after some thought, the children saw fit to qualify this method with 'as long as the friend knows how'. The children could tell the difference between doing a job using what you know and doing a job using guess-work.

Some of the concept learning was by solving problems of moving joints by trial and error. The children welcomed the opportunities to put technological skills to use and this method of learning gradually increased in acceptance the more they acquired the skills they needed. I encouraged a supportive atmosphere during trial and error sessions, to enable willingness among the children to regard the process as a means of learning, even though things often did not go according to plan. The younger children were particularly vulnerable in this respect, and I sought guidance from Elsie Osborne's book, *Understanding your 7 year old*. In it, she stated:

> Confidence may ... swing between too much certainty, based on new-found skills and too little when performance does not come up to expectations. This is a time ... when children can be very self-critical and also sensitive to other people's comments.
>
> (1993: 23)

The class atmosphere had a great bearing on the quality of learning and in the first few weeks it seemed as if there would never be a time when I could look forward to a trouble-free lesson. The children, however, proved to be supportive of my aims to develop a sense of pride in finishing work, with give and take in cooperative working, and concentrating on efforts to think in terms of design and making tasks. There were far fewer disruptive episodes during the last few weeks and the children concerned appeared to be becoming interested in their work, through their own involvement and the tolerance of their peers.

I found that the self-esteem of these children was showing signs of improving so I gave individual help at the start of a session, and frequent general encouragement to succeed by finding something to praise, however small. Their concentration spans improved sufficiently for them to remember past work and build on it. The children must have valued my one-to-one teaching for it to receive almost half the vote points. I found myself thinking again of Holt's claim that 'memory works best when unforced, that it is not a mule that can be made to walk by beating it' (Holt 1991).

On the whole, I thought that the children had made good choices of preferred learning methods. They were appropriate in that they fulfilled the majority of their needs. I wondered what their preferred methods of learning would be for other subjects, recognizing: 'The importance of getting the pupil to participate, and to make the educational situation 'come alive' (Schools Council 1983: 104).

SUMMING UP

My initial concern was to find out how to deal with the situation of teaching younger children, including ten with behavioural patterns of a disruptive nature and four who needed one-to-one adult help. This prompted me to investigate the learning that took place related to Jointed Movement in design and technology. The results of my enquiry showed me beyond doubt that by teaching the children in ways which they preferred and responded positively to created a successful learning environment, both in 'effectiveness' and social terms. There was also sufficient evidence to challenge my own role as teacher as I had previously assumed it. The experience taught me that I was too restricting in my approach, and by deciding to examine my contributions to the children's learning I saw the reasons for change.

The learning opportunities that I offered on a regular basis did include the children's favourite methods, but not in proportion to their needs or preferences. On reflection, my main fault appeared to be in not withdrawing myself sufficiently for them to embark on their own peer learning. When I did allow for this to happen, I was heartened to find that disruptive behaviour was minimal, and all the children were motivated to contribute in a positive way without relying on me for support. It became apparent that there was a unifying force at work which stemmed from the children's own interest. They were involved with applying conceptual understanding, in cooperation with each other, and I was able to 'take a back seat' for far longer than I had anticipated.

My final evaluation was the culmination of these weeks of making use of the range of learning opportunities, and it resulted in my list of 'methods' being revised. The following list shows these in order of priority as influenced by the children's reactions:

- going on special outings, reading books, looking at pictures, watching television, doing own research as stimulus for ideas;
- experimenting and playing with materials;
- children helping each other, partners/groups;
- design and making through trial and error;
- Children discussing work together while working as individuals;
- encouraging the children to ask spontaneous questions that enhance their understanding;
- evaluation of work done shared within and outside school; making a display for use as a teaching aid;
- step-by-step instructions whether spoken, written, drawn or demonstrated; blackboard use, as required;
- teacher helping child on one-to-one basis; also supervising for safety;
- selection of work for 'Good Work' Assembly; praise for effort as well as for quality;
- linking a task to a real-life situation;
- review of work done over a period of time.

I remember starting out with my class with much apprehension. I was relieved to find that my own learning increased, not only conceptually but also in terms of give-and-take in response to the needs of myself and the children. The children's parents were also knowledgeable regards our work, suggesting that the children were talking about it; my next task will be to build on this work by giving appropriate learning experiences for the design and making of powered models. In the climate already created, it should be a rewarding experience.

FURTHER READING

Department for Education (DfE) (1995) *Technology in the National Curriculum*, London: HMSO.
Department of Education and Science (DES) (1987) *Curriculum Matters 9*, London: HMSO.
Holt, J. (1991) *How Children Learn* (2nd edn), Harmondsworth: Penguin.
Osborne, E. (1993) *Understanding your 7 year old*, London: Rosendale Press.
Schools Council (1983) *Working Paper 75, Primary Practice*, London: Methuen.
Soper, S. (1987) *Primary First*, Oxford: Oxford University Press.

Chapter 17

Positive discrimination
Is there a case?

Dene Zarins

There had been much discussion in our staff room of late, about equal opportunities and the use of construction toys. Some of us had attended a course at another first school, where Chris Brown had suggested positive discrimination in favour of the girls (see Brown 1990). Of course, we said, it's obvious! Boys are much keener and more able to use and construct models. Given the highly motivated and articulate nature of many of our girls, why should this be so? The results of an initial survey indicated that many of the girls had experience of using construction toys at home. Indeed, the girls seemed to share skills and motivation equally with the boys when engaged in directed technology tasks (such as card junk, woodwork and clay models). Inequality, though, was an idea that I was aware of, as we worked through the following weeks. There were two things in particular about this sequence of events that made me stop and think:

- In the first one and a half weeks that we were using the newly acquired giant construction kit, all the users had been boys. Why boys only? The boys and girls had equal opportunities to use the kit. Was it that the boys had more confidence? Did they regard construction toys as 'boys' toys'? Did the girls perceive this gender bias too? Once settled to their task with the kit, did the boys dominate the space and equipment, thus excluding the girls?
- I should have been aware that initially there were no mixed groups or girls' groups playing with the kit. Was the use of the kit high on my list of values? Was its use valued by the girls? The other girls didn't start to use the kit until they saw girls receiving praise and positive feedback from their teacher and peers. Did the girls lack confidence with construction toys? Were their expectations of their capabilities low? Has their past experience in using the construction toys not helped them build their confidence?

My class has sixteen Y1's and sixteen Y2's; fifteen boys and seventeen girls. There has not been much research done regarding girls, technology, and construction toys in the early years of schooling. Christine Brown, in her article 'Girls, boys and technology' has said that 'information about what girls can do in the pre-secondary stage is still very sparse' (Brown 1990: 2). Her own research set out to explore that issue, using construction equipment such as 'Lego' in teacher-directed activities, and suggested that girls make different kinds of models from boys. Other observations, made in Ealing first schools, have shown that in free-choice sessions construction is generally a boys' domain and that girls choose construction less frequently and do not stick at it (Claire 1992). Could it be that I was observing a 'mirroring' of these attitudes when I noticed that the girls seemed 'less keen' to choose construction toys at the age of five? If so, how important is it? Some research has shown the importance of construction experiences at an early age. Kelly, for example, has pointed out that children who play with construction toys and handle tools probably develop both their spatial ability and their scientific aptitude in the process (Kelly 1987). If there is evidence that girls have a negative attitude towards technology, and it is also shown that construction work is potentially a valuable aid to the children's learning, it would be wise to find the extent of the negative attitudes and seek ways to combat them within my own teaching.

I decided to look at the use of the construction toys during a free choice session, and see if the children who used them were indeed predominantly boys. I decided to use a 'fly on the wall' technique, to note down everything that went on in the carpet area (where the construction toys are kept) initially for one session (75 minutes). This initial investigation gave me more information than I expected. Two boys and three girls used the construction toys, but it was the way they were using them that drew my attention. The boys worked together on making a model, testing it and improving it, while the girls set up the Playmobile people and hospital and used it as a 'jumping off' point for creative play. Following this observation there were now two strands that I wanted to follow:

- Did more boys than girls choose to use the construction toys? This could be quantified by noting down numbers of boys and girls using the toys over a period of weeks.
- What was the quality and type of play that was going on? This could be recorded in various ways, e.g. by writing notes, videoing a construction session, and taking photographs.

For the purposes of this research I planned that each classroom group should have the opportunity to choose construction toys during each week, and that I would record their use over a four week period.

In the four weeks 73 of the users had been girls, and 93 boys. In my planning I had made sure that each group had equal access to the construction toys. Perhaps because of this there had been more interest shown in

Table 17.1 Numbers of boys and girls using the construction toys

	am		pm	
	Boys	Girls	Boys	Girls
Week 1				
Monday	3	2	4	4
Tuesday	2	2	3	0
Wednesday	4	1	3	2
Thursday	2	3	4	2
Friday	1	0	5	4
Total: 31 boys and 20 girls.				
Week 2				
Monday	2	3	4	4
Tuesday	0	0	1	0
Wednesday	0	0	2	1
Thursday	3	2	2	3
Friday	0	0	0	0
Total: 14 boys and 13 girls.				
Week 3				
Monday	(staff training day)			
Tuesday	0	0	3	1
Wednesday	2	0	4	3
Thursday	2	2	0	0
Friday	3	4	8	4
Total: 22 boys and 14 girls. The numbers were not balancing out as I had anticipated/hoped.				
Week 4				
Monday	0	0	2	3
Tuesday	0	0	4	4
Wednesday	2	4	3	1
Thursday	4	4	2	4
Friday	3	2	6	4
Total: 26 boys and 26 girls.				

these activities by both sexes, with opportunities for choosing the toys being the same for both. By 'equal access' though, I had thought in terms of equal time and opportunity to use the construction toys. Reading through Hilary Claire's article made me think again. She found in her observations that the boys had assumed that the space and resources were theirs by right (Claire 1992). This made me wonder if I am really planning for 'equal access' when allowing mixed sex groups to use the equipment, given that the boys more 'robust' behaviour and use of space may put the girls off choosing the same activity.

I had initially assumed, or at least hoped, that there would be a fairly even number of boys and girls using the equipment through each week. This initial set of observations showed my hope was ill founded even when I consciously planned for equal access. I now began to ask myself if this discrepancy in numbers was enough on its own to warrant positive discrimination in the use of construction toys, in favour of the girls. However, the observation records were also too simple. They did not tell me, for example, if the children worked in mixed or single sex groups. I needed to get more information about this. Perhaps a video of a typical construction session, or still photographs, would record the data I wanted.

During a morning session three groups of children were working on maths and language activities. The fourth group (Year 1 children) had been given time to use the carpet area, and were asked to choose one of the activities there: large construction kit; 'Lasy' construction; Lego; Polydrons; 3D shapes; Fuzzy Felts; magnetic picture kit; jigsaws; Playmobile set; connecting figures; threading beads.

What I was looking for was evidence of who the children chose to work with; their choice of construction toy; how they worked with each other; how they sustained the activity; what skills they used; and their use of role play. I recorded a transcript of the video in the form of notes, detailing the moments as I interpreted them, like an observation diary. The interpretation of events is indicated in the notes:

Lucy, Nathalie and Sophie chose together. Nathalie makes the choice of magnetic pictures, Lucy and Sophie watch. Kyle, Sam and Simon all choose Lasy. They pick up an almost completed model and connect it to the motor. Kyle works on own model, joins in with Sam and Simon occasionally.

Girls are negotiating each stage of their picture. Four more boys come in, glance at girls' picture and home in on Sam and Simon's model. 'That's good', said Michael. Teacher moves that group back to their activities.

Simon tries out the 'helicopter' model. Lucy leaves. Kyle finishes his model. 'Simon, this is what I've done!' he says and tries to show it to him. Simon is involved with Sam. Kyle tries again. He 'flies' his model. George comes to watch.

Sophie and Nathalie finish their picture and turn it upside down, chant '1,2,3!' and shake all the bits off. Kyle glances over at the girls, and tries to attract Simon's attention again. Simon looks at Kyle's model. Kyle explains what it does 'then it flies ... and that ...' and shows him.

Nathalie picks up Lucy's magnetic picture and tips it up. Both girls start to clear it away. Kyle looks over. Nathalie sits on the box and begins to throw bits in. Sophie talks to Kyle and Nathalie watches.

Positive discrimination 205

Figure 17.1 Girls role playing using Playmobile, boys working with Lasy

Lucy moves in to clear away. Kyle says to the girls 'I've got a thing in my pocket', Simon moves over. 'Can I help?' says Kyle to the girls, 'Why don't we help . . . with this one?'

Sam takes a look at the picture and moves back to his model. Nathalie and Sophie look in the shelves. Nathalie picks up the 3D shapes and Sophie a book. Sophie reads and Nathalie puts the shapes back. Nathalie picks up the Fuzzy Felt and puts it back, plays with her hair, and does a little dance, Sophie dances too. Both look into the Lasy box and jump back when Alex comes in.

Sam shows him his model '. . . and then we made this . . .' the boys all move over to the Lasy box. Nathalie and Sophie tip over one of the boxes onto Kyle's model. 'That's mine!' he says and picks it up. Nathalie and Sophie pick up the bits and put them back into the box.

Kyle reconnects the helicopter model and plays with it. Sam and Simon work together on Sam's model. Nathalie throws more Lasy over the box. 'Can I have a go?' Nathalie asks and takes the model from Kyle. Then Sophie holds it. Kyle keeps hold of the battery holder. The girls hold the helicopter under their chins. Nathalie gives it to Simon and she and Sophie move to their seats, where they play about for a while.

Figure 17.2 George and Simon trying to get the motor going

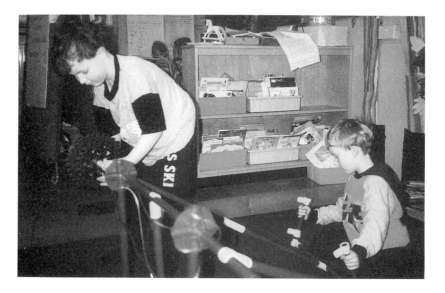

Figure 17.3 Kyle is adjusting the drive machine, Sam is driving it

Figure 17.4 Sam and Simon driving their machines

As the children moved into the carpet area they split into two distinct groups, boys and girls. When the boys approached the girls and asked if they could help, the girls withdrew from the area. The girls approached the boys briefly, but did not sustain any activity with them. When the girls had finished their first activity, one left the area for about twenty minutes, while the other two girls stayed on, together. The three boys stayed together, for the hour. In summary, this detailed observation showed me that:

- the children split into boy/girl groups when choosing construction toys;
- some construction toys offer a greater range of open-ended activities than others;
- the boys chose a construction toy that offered more opportunities for a wider range of activities than the girls did;
- the boys sustained their activity for a longer period;
- the boys made their models, tried them out and adapted them;
- the girls moved quickly to an end-product, abandoned it, and moved on;
- both boys and girls worked collaboratively with their own sex, helping, by offering each other more apparatus, admiring each other's models, and suggesting ways to improve them;
- there was little interaction between the sexes;
- there was a difference in the use of role play.

This observation had not told me anything about;

- why the children split into boy/girl groups;
- why they made their choice of toy;
- why the boys sustained their interest for longer;
- why the girls did not sustain their interest;
- why the difference in role play occurred.

I decided to take photographs of children, during typical working periods, recording what they chose during a 'free choice' session. In addition to the carpet area options, desk top activities were available: book making, picture making, message writing, code making. Playing in the shop, in the sand pit, or working on the computer were also options. Of the six children in the group photographed, three girls and one boy chose paper and pencil activities, one boy chose the computer, and one girl chose a construction toy on the carpet.

From these photographs it was clear again that the boys and the girls were working in separate groups, each working cooperatively. It is also clear that the visitors and watchers in the carpet area, during these sessions (this is borne out by the video), were more often boys. Given this evidence of 'incidental learning', it may be that I need to give more attention to who the 'watchers' are and what they may be observing.

In puzzling about the reasons why the gender split was so evident, I wondered if there might be a link between the children's choice of 'free choice' activities and their experience of construction toys at home? If it is correct that children who play with construction toys and tools probably develop both their spatial awareness and their technological aptitude in the process (Kelly 1987) then their pre-school/home experience might already have played a part in bringing about the differences. Could it be that lack of experience of construction toys leads to lack of confidence in this area in the classroom? Earlier I asked, when talking about girls' negative attitudes to technology,

> How far back do these feelings develop? Could it be from the age they are given toys to play with? It has been said that 'conventions of child rearing today ensure that such toys and tools are much more frequently made available to boys than to girls'.
>
> (Kelly 1987: 2)

I decided at this stage to find out from my own class what sort of construction toys they had, and played with, at home. First of all, we talked together about the construction toys that we had available in the classroom, and together we made a list. I told the children that I would ask them if they had any of these toys at home. I also asked them if they considered each toy to be a toy that was suitable for girls and boys, boys,

Table 17.2 The childrens' responses to the questions: which of these toys do you have at home, that you play with? Are these toys both girls' and boys' toys, boys' toys, or girls' toys?

Toys played with at home			Girls' and boys' toys	Boys' toys	Girls' toys
Toys	Boys	Girls			
Sand	6	2	29	0	0
Water	10	11	26	2	1
Easyform	5	7	24	5	0
Puzzles	11	6	29	0	0
Giant Constr.	3	0	7	22	0
Lego	11	10	12	15	2
Polydrons	3	3	10	11	8
Art-straws	4	5	8	12	9
Stickle Bricks	3	0	12	7	10
Lasy	2	0	8	19	2
Georello	0	0	6	5	20

or girls. In order that they were not swayed by peer group pressure, I asked each child individually, and requested that they kept their answer a secret until the end of the session. There were twenty-nine children present at the time.

Table 17.2 shows that both the boys and girls do have experience of construction toys at home, but while looking at it, some points have to be taken into consideration. The first is the problem of carrying out research with young children. It was difficult to know how dependable this evidence was. I was suspicious, for instance, that there were some toys, such as the giant construction set, and the Lasy, that the children would like to have had, and therefore claimed ownership of, at home. Conversely, some children might not claim ownership of or admit playing with Stickle Bricks as they regarded them to be 'babyish'. With that proviso in mind, a simple totalling of 'instances' of owning and using such toys shows there were 80 instances of boys and 65 instances of girls. To my mind, this may not be significant enough a difference to justify concern. It was when examining the responses to the questions: Is this a toy for both boys and girls? Is this a boys' toy? Is this a girls' toy? that there were some disconcerting responses. Whereas most of the children considered sand, water, Easyform and puzzles were for both boys and girls, when it came to the giant construction kit, twenty-two of the children considered it to be a boys' toy. This was a response that I wanted more information on, so I asked some of the girls why they had felt that about the kit. One of the girls said: 'The construction set is for boys because they play with it more.' Another added 'boys are stronger, the kit's hard to use'. The boys' responses were similar: 'Girls don't play with it that much.' One also said: 'There are more things you can make with it that boys like, like cars or aeroplanes' and 'Boys like building big things'.

While talking to the girls about this they seemed reticent and ill at ease, even unwilling to give their reasons as to why they considered this a boys' toy. Perhaps they had picked up on my interest, maybe they didn't like 'admitting' that an activity was not for them, or it may have been that recognizing that there were 'boys' toys' lowered their standing in the eyes of the boys. The boys, however, were relaxed about giving their responses and showed none of the reticence of the girls. They may have been more confident with the issue and felt that the answers they gave did not undermine their standing with the girls.

Given this evidence of the children's views, it is hardly surprising that fewer girls choose to use the construction toys than boys. One question which therefore arose was: If the girls chose not to use the construction toys, then how would they fare, given a teacher-directed technology challenge which involved such use?

I decided to give the children such a task, and look at the way they responded to it, to see if there was a detectable difference in the boys' and girls' responses. The three Year 1 and 2 teachers usually share the planning of topics, and the one that had been agreed upon for the Spring term was 'Giants'. We had been asked to plan a technology project concentrating on an area that we did not feel confident in. To that end, as well as using card, and the contents of the junk box, we had wood and tools available for the children. The starting point for our topic was based on the story 'Jack and the Beanstalk'. From the story there are many opportunities for technology based activities. We converted some of them into design and make tasks:

- Jack has to get the hen that lays the golden eggs to the beanstalk quickly. Design a wheeled carrier to carry the hen.
- How can Jack carry the hen safely down the beanstalk? Design a way for him to do this.
- In the story the magic harp cries out for help. Make a harp of your own that will make a sound.
- In the sand pit make a model/map of what the giant can see as he looks down from the beanstalk.
- Using Lego, make a wall to go round the giant's castle. It must be strong enough to keep other giants out.
- With large construction apparatus – make a giant figure.

Junk box contents, elastic bands, paper fasteners, card, cotton reels, wheels, wood (assorted), balsa, dowelling, glue, balsa cement, masking tape, polyform, paints, string, buttons, drawing pins, could all be used. Scissors, hole punch, drill (on stand), G clamps, hacksaws, and other small handtools suitable for early years pupils can be used. The skills involved were identified as including: drawing, oral language, writing, measuring (and seeing the need for it!), cutting card and wood, joining, modelling, adapting plans, evaluating.

I started the technology project with one Year 2 group, while a co-teacher worked with the other groups. I talked about the story that we had been reading. I mentioned that Jack stole the hen from the giant's house, and we talked about how he carried it to the beanstalk. They suggested: 'He put it under his jumper'; 'in his hat'; 'tied a bit of string round its neck'. I asked them if they could design a carrier with wheels, but while they were thinking about it, I asked them to look at the materials and tools. I also showed them how to cut the wood and how to make joins. We talked about making a design first and what we would need to put on it. Some children chose to work together and some alone. Blue Group started straight away. The children were desperate to get started and rushed their designs so they could get their hands on the materials. Mostly, their pictures showed the general shape of the carrier, though they were not sure how they would fit together the sections of materials and components.

Rahma, Hannah and Emma worked at producing a box shape (to carry the hen). Michael and Mark made a box with long legs with cotton reel wheels fixed on with elastic axles. William and Philippa made a square and were thinking of ways to make sides. Jamie drew his plan as a series of unconnected parts. He said he had a clear idea of what he was going to do and how he was going to do it. The children were enthused by the idea of making a hen carrier and by getting their hands on the wood and the tools. There was much handling of the wood, and discussion about shape and size, and speculation as to whether it would work before they began to cut the wood.

All eight children completed the challenge to design a carrier that will take the hen that lays the golden eggs from the castle to the top of the beanstalk, from the design stage, to suggesting what they might do to improve their model if they had another go. I could not detect a difference in the ways the boys and girls tackled this task. The girls and the boys responded equally to it, in terms of the ideas, problem solving, and skills used.

Initially, looking at my own class and how the girls operated, being articulate and assertive, I thought there was no case for positive discrimination in their favour. Of course, I tried to make sure they had equal access to the construction materials and equipment, and the nature of the task was identical. This was not a free-choice situation. I began to see there were more issues than I had realized in relation to the activities in the carpet area, and the use of construction toys during 'choosing' or 'free choice' time. In those situations I was certainly giving the children equal time to use the construction toys. Yet by planning for mixed groupings

I was not creating the opportunity for the girls to choose the construction toys. They felt unable or unwilling to do so, given the boys' dominant use of space and both the boys' and the girls' gender bias in their perceptions of construction activities.

I prefer to have mixed grouping in my teaching and try to make sure that I don't segregate the children by sex. I feel that boys and girls should learn to work and share together. It came as a jolt to realize that there may be some areas where teacher directed activities and single sex grouping works better, and that construction activities might be one of them.

This research evidence is not to be seen as in any way providing a basis for conclusive statements. However, certain questions have emerged that I would wish to address in terms of my own practice, and as such, they are presented as questions that may also be applicable to other teachers:

- Do I provide/create opportunities for girls to experiment and work in supportive girls' groups?
- Is there adequate space to use the kits?
- Do I monitor and guide the girls back on task if they are stuck?
- Do I provide enough direct and supportive teaching in the use of construction toys?
- Do I give enough praise and positive feedback, and encourage the same from my children?

As a result of this initial research, and a new awareness of the issues involved, I intend to work towards addressing all of these questions in an attempt to encourage girls to choose and be enthusiastic about using construction toys equipment, and materials. To develop the skills that will enable them to become happy and confident with technology tasks it seems I will need to manage situations more 'intentionally'.

FURTHER READING

Brown, C. (1990) 'Girls, boys and technology', *School Science Review* 30 (7): 1–9.

Claire, H. (1992) 'Interaction between girls and boys: working with construction apparatus in first school classrooms', *Design and Technology Teaching* 24 (2): 25–34.

Kelly, A. (1987) 'Science for girls', *HMI Series: Matters for Discussion – Girls and Science*, London: HMSO.

Lord, W. (1987) *The Nature of School Technology – Technology Education Project*, London: St. Williams Foundation/HMSO.

Smail, B. (1984) *Girl Friendly Science: Avoiding the Sex Bias in the Curriculum*, Harlow: Longman.

Index

ability levels 50, 63, 92, 112, 166, 178
action research 11–13, 14–15, 16; classroom reality 121; overview 125; qualitative methods 19–20; quality of 26–8
Andersen, L. W. 125
Arnheim, R. 96
Assessment of Performance Unit, DfE 3

Bassey, M. 121
Brittain, W. L. 95, 96, 107, 109
Brown, C. 201, 202
Burns, R. B. 125

case study research 12
celebration, food for 169–70
Child, D. 81
children: conceptual awareness 178, 179, 185, 187, 194–5; confidence 198; cooperation 62, 80–1, 83, 112, 116, 119, 189; copying 80, 81–2; evaluation 49, 65–6, 69, 122–3, 149, 154, 161–2, 165; gender differences, construction kits 201, 202, 203, 208–10; independence 196, 197; interaction with teacher 49, 63–4, 93, 111; mixed ability 50, 63, 92, 112, 166, 178; reflective/impulsive 163–5; understanding 179, 185, 187, 194–5
Christmas card project 146, 150–1
Claire, H. 202, 203
classroom 18, 121, 198
Cohen, L. 80–1
collage calendar project 146, 147–9
common ownership 84
communication skills 81–2, 196, 197

communications project 41
competitiveness 83
confidence: children 198; teachers 77, 111, 132–3, 141–2
confidentiality 17
Conger, J. 163, 164
construction kits: availability 134, 135–9, 140, 142; as boys' domain 202, 203; gender differences in use 201, 202, 203, 208–10; home use 208–9
cooperation: children's perception 62; and copying 80–1, 83; group work 112, 116, 119, 189
coordinators 11, 30, 132
copying 77–8; children's views 80, 81–2; collage 147; crediting originator 82–3; legitimate 80; and permission 80, 81–2
copyright 78
creativity 81, 83; and group work 84–5; as social currency 83–4; teaching styles 78–9
Croll, P. 78–9
cutting skills 44, 46

data 17–20; analysis 23–4; children's logs 113, 116; collection/interpretation 124–5; design and make tasks 51–2; distortion 24; interpretation 23; organisation 22; recording 18–19, 22, 113, 121; *see also* observation
Dearing Review 2, 4
DES 3; (1987) 178; (1990) 157; (1992) 129
design brief 163, 165, 166, 176–7
Design Council 36, 120

design and make tasks 4; ability levels 166; for assessment 159–60; completion 39, 46; evaluation 52, 56–7, 162–3; girls/boys 210–12; in use 30, 34, 51–2

design and technology 36, 45; coordinator's job 11, 30, 132; cross-curricular 30, 168, 178; drawing 95–6, 97–103, 106–7, 109, 114–15; end-product 104–6, 115; as homecraft 88; imaging/realizing 95, 100, 102–3, 104–6, 107–8; inservice training 119, 142, 143; mixed ability children 50, 63, 112, 166, 178; national curriculum 1–4, 8, 66, 95, 120, 131–2, 156, 158, 159; problem solving 3, 4, 78–9, 120, 125, 140, 156–7; progression 131–2, 134, 137–9, 140; quality 46–8, 49; resources 38, 39, 77, 132, 133–4; skills taught 40, 78–9, 120, 137–9, 140, 196; teachers' confidence 77, 111, 132–3, 141–2; timetabling 38, 159; as woodworking 40, 42, 46; *see also* design and make tasks; materials

design and technology projects: alarm system 41; bridge 31; Christmas card 146, 150–1; Christmas tie 31–2; collage calendar 146, 147–9; communications 41, 43–9; flight 87–93; food for celebration 134, 135–9, 140, 168–77; furniture of the future 51–9, 62, 63; jointed movements 178–98; milk carrier 122–3; puppet theatre 146, 151–3; *shaduf* 113–18; Titanic 33–6; vehicles 79–80, 99

designer, capabilities 122, 125, 126–9

designing 120; Attainment Targets 161; design and technology 39–40, 43, 44, 51, 52–6, 63; hidden plan 102

DfE: (1995) 2, 4, 8, 65, 66, 95, 158, 159; Assessment of Performance Unit 3

Dodd, T. 78

Downey, M. 84, 112

drawing: as *aide-memoire* 99, 100–1, 109; annotated sketches 97, 98, 99, 109–10; content and elements 106–9; for design and technology 95–6, 97–103, 106–7, 109, 114–15; details 106–9; developmental stages 96; for general outlines 101–2; improvements 191; and making 100, 102–3; representational 96, 107; scale ratios 107; as starting point 104–6; styles 96; technical 107–8, 109; users 103–4, 109; uses 99–103

Education Act (1993) 2
Education Reform Act (1988) 2
Eggleston, J. 4, 156
electrical components 31–2, 69, 72
end-products 104–6, 115
evaluation: appearance 67, 68, 70, 72–3, 74–5; as attainment target 157–8, 159–60; children's 49, 65–6, 69, 122–3, 149, 154, 161–2, 165; design and make tasks 52, 56–7, 162–3; function 67, 68, 70, 73, 74; gender differences 69; group work 117–18; improvements 165–6; learning 186–7; manufacture 67, 68, 70, 73, 75; manufactured toys 65–70; own work 69–71, 76; social expectations 71; teacher's 117–18, 161–2; *see also* self-evaluation
executive process 157

failure, as learning factor 71
flight project 87–93
food: for celebration 169–70; in design and technology 134, 135, 140, 168–77; ready-made 171, 176
Freeman, N. 95, 96
furniture of the future project 51–3; design brief 62, 63; designing 51, 52–6; evaluation 52, 56–7; model making 57–9

Galton, M. 78–9
Gardner, H. 95, 96, 107, 109
gender differences: construction kits 201, 202, 203, 208–10; evaluation 69
glueing 44, 58
Goldscheider, L. 110
Goodson, I. F. 3
group work: cooperation 112, 116, 119, 189; creativity 84–5; discussion 151; drawings for design 114–15;

evaluation 117–18; individual input 84–5, 114; making stage 115; mixed ability children 112; negotiation 116–17; organization 112; partner selection 113–14, 121, 123–4, 126–8; personality clashes 114, 129; reflective/impulsive children 164–5, 166; single sex 43, 212; skills 129; vertical groups 120, 197; working relationships 116

health and safety 170
Hennessy, S. 156–7, 162, 165
Hirst, P. 36
Holt, J. 197, 198
homecraft 88
How Children Learn (Holt) 197

inservice training 119, 142, 143
intellectual property 78, 83

Jeffrey J.R. 86
Jinks frame 69, 72, 79
Johnsey, R. 50, 120, 156, 160
jointed movement project: assessment 191–2; ball and socket 184–5; conveyor belt 188–9, 191; evaluation 186–7; folding and bending 180–1; hinge joints 181–2; hydraulics and pneumatics 182–3; learning methods 183–4, 192, 196, 198; learning opportunities 179–80, 195–8; shopping trolley/hand cart 189–90; supermarket visit 190–1; swivel chair 187–8; worksheets 193–5

Kagan, J. 157, 163–4
Kelly, A. 202, 208
Kelly, A. V. 36, 112
Kimbell, R. 106, 109
knowledge perspectives 35–6

learning: building-block method 171; cross-curricular 36; through failure 71; from friends 61, 80, 197; guidance 154–5; incidental 208; methods 183–4, 192, 196, 198; opportunities 179–80, 195–8; personal meaning 93; social setting 83; trial and error 189, 190, 197; use of 178–9; worksheet evidence 193–4
learning environment 198
Lees, J. 63
literature, for research 21
logbooks 113, 116
Lowenfeld, V. 95, 96, 107, 109
Lund, D. 63

McCormick, R. 156–7, 162, 165
making skills 80, 158
making stage: and designing 115, 211; evaluating 165; importance 39–40, 43; models 57–9
Manion, L. 80–1
manufactured products 165, 171, 176
materials: access 89, 140, 155; evaluated 65, 132, 133–4; recycled/construction kits 134, 135–9, 140; selection 90–1, 109, 114, 150–1, 152
milk carrier project 122–3
models: making stage 57–9; mechanisms added 59–61; playing with 43–4
Mussen, P. 163, 164

national curriculum, design and technology 1–4, 8, 66, 95, 120, 131–2, 156, 158, 159
National Curriculum Council 131–2
negotiation, group work 116–17
Newton, D. P. 50
Newton, L. D. 50
Nuffield Science cards 33

observation: in classroom 43–9, 88, 113; construction kit play 202–8; problem solving 159–61
open-ended task: design 92; discussion 88–90; end product/processes 63, 92; evaluation 91, 93; food 170, 171; mixed abilities 50, 92; planning 172–5, 177; teacher input 87–8
open-mindedness 16, 28
openness, in research 16, 28
Osborn, E. 198
outcomes 86, 145, 153
ownership: in common 84; drawing 103–4; intellectual 78, 83; and kits 140, 142

Parkes, M. 4
partners for work: *see* group work
Pearson, L. 157, 164
peer groups 112
peer help 61, 80, 197
planning 100, 147–9, 155; *see also* designing
play 202–3, 207
Primary Practice, a Sequel to the Practical Curriculum 195–6
Pring, R. 36
problem solving: cross-curricular 120; national curriculum 3–4, 78–9, 120, 156–7; skills needed 125, 140
Problem Solving in School Science (Johnsey) 156
product-led tasks 170–1
professional development 10, 179
project packs 142
prototypes 51, 53–6, 58, 89, 115
puppet theatre project 146, 151–3

quality: design and technology 46–8, 49; in research 27
questionnaire 42, 67, 88, 161, 165
questions, for research 125

Raush, H. L. 121
recycled materials 134, 135–9, 140
references 25
reflection–impulsivity 163–5
research, types: action 11–16, 19–20, 26–8, 121, 125; case study 12; qualitative/quantitative 19; teacher 5, 10, 12–13, 27
research projects: action plans 24; background reading 21–2; beginning 13–14; data 17–20; designing 22–3; negotiation and interaction 16–17; and principles 16–17; questions 125; references 25; resources 20; topics and focus 14–15
research report 25–6, 27
resources 38, 39, 77, 132, 133–4; *see also* materials
role play, construction kit 202–3, 207
rubber-band powered vehicles 79–80

scale ratios 107
Schools Council 119–20, 195–6, 198
Schools Curriculum and Assessment Authority (1995) 2–3

science, and design and technology 32–5
self-esteem 198
self-evaluation 69–71, 76, 161, 162
self-expression, language/drawing 96
shaduf working model project 113–18
Simon, B. 78–9
skills: communication 81–2, 196, 197; group work 129; for making 80, 120, 137–9, 140, 158; problem solving 125, 140; study skills 196; tools 40; training 78–9; woodworking 42, 44, 46
spot-skill teaching 167
staff development 10, 11
Stanford, G. 112

tasks: completion 69; product-led 170; and replication 86–7; *see also* design and make tasks; open-ended tasks
teacher: confidence about design and technology 77, 111, 132–3, 141–2; as demonstrator 180; as evaluator 49, 104; as helper 43; interaction with children 49, 63–4, 93, 111; role 86–7, 196; as safety supervisor 180; as skill demonstrator 171, 176; as standard raiser 147
teacher consultant 119
teacher intervention 153–4, 166
teacher research 5, 10, 12–13, 27
teacher-practice 178
teaching outcomes 86, 145, 153
teaching strategies 145, 167–8
teaching styles 78–9, 86–7, 167
technical drawing 107–8, 109
technology: *see* design and technology
Technology Working Party report 4
textiles 135–9
Tickle, L. 3, 5, 27, 27, 153–4
time: as resource 20, 32, 40, 41, 42–3, 44, 69, 79; restrictions 142
timetabling, design and technology 38, 159
tools 40, 43, 58
toys 65–70

understanding, children's 179, 185, 187, 194–5
Understanding your 7 year old (Osborn) 198

values 65–6, 73
vehicles: for Lego person, design task 99; rubber-band powered 79–80
vertical grouping 120, 197
views 65–6, 73

Welch, L. 157, 164
Willens, E. P. 121
wood, cutting skills 44, 46
woodworking 40, 42, 46
work groups: *see* group work
worksheets 193–5